the ODD BODY

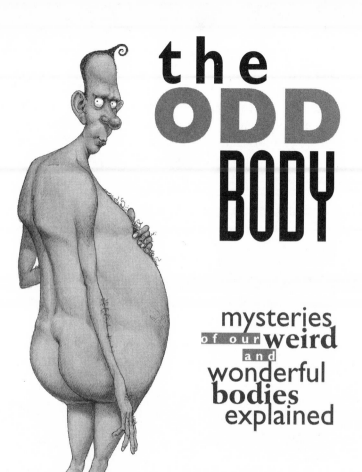

the
ODD
BODY

mysteries
of our **weird**
and
wonderful
bodies
explained

Dr. Stephen Juan

**Andrews McMeel
Publishing**

Kansas City

04 05 06 07 08 MLT 10 9 8 7 6 5 4 3 2 1

ISBN: 0-7407-4188-8

Library of Congress Control Number: 2004101696

Illustrations by Rod Clement

ATTENTION: SCHOOLS AND BUSINESSES

Andrews McMeel books are available at quantity discounts with bulk purchase for educational, business, or sales promotional use. For information, please write to: Special Sales Department, Andrews McMeel Publishing, 4520 Main Street, Kansas City, Missouri 64111.

To Ken, Marti, and C.J.

contents

acknowledgments

There are many people to thank for making this book possible. At the top of this list are the scientists and researchers who investigate the human body and how it works. Some are from the medical sciences, some from the behavioral sciences, and some from other fields. This book would have nothing to say if there were no body of knowledge to communicate.

Next are the curious people who ask questions, especially those who can't sit still until they get the answers. Many of these are my own university students, who have asked me questions about "us" that I have always tried to answer as best I could. Curiosity may have killed the cat, but it also leads to knowledge. Thus, it is always to be encouraged. And I want to thank too those individuals whose words have sparked my own curiosity.

Thanks are also due to the librarians at the University of Sydney, who make resources available and who are always cooperative and helpful.

Thanks too to the people in Australia who encouraged me to undertake this project: Mel Cox, Melissa Gabbott, Angelo Loukakis, Jude McGee, Lisa Mills, and Linda Tenkate.

Thanks also to Rod Clement for his fantastic illustrations.

Thanks to my wife, Buffy, and to our two daughters, Alicia and Cassie, who gave me time and space to produce whatever of value is between these pages.

Thanks are also owed to the people who continue to urge me to write more books (although I need no urging): Airlie Lawson, Fiona McLennan, and Alison Urquhart.

Many thanks to my agent, Stephen Ruwe, for all the help he has given me and the effort he has made to get my works more prominently to the U.S. audience.

Thanks finally to the people at Andrews McMeel Universal for publishing this revised and enlarged edition of the first of my Odd Books: Jennifer Fox, Peggy Hoover, Pete Lippincott, John McMeel, Kathleen Andrews, Tom Thornton, Chris Schillig, Hugh Andrews, and Dan Boston.

intro**duction**

Has this ever happened to you?

Did you ever want to know something about the human body but were afraid to ask? Or you didn't know whom to ask? Or there was nobody around to ask? Say you wanted to know why you yawn, why your skin wrinkles after a bath, something silly like why men have nipples, or something really weird like can you keep a severed head alive?

At home you may have thought about asking your parents. Maybe you even tried. But more often than not they didn't know themselves. If they told you to look it up (the face-saving suggestion to maintain a parent's dignity when faced with their own ignorance) and you did, you probably couldn't find a book that gave you the answer you wanted. So you moved the question to the back of your mind and eventually forgot about it. A few years later, in biology/life education classes at school, the question may have occurred to you again. Should you ask the teacher? You decide not to. After all, the question was off the subject, it would take up class time, your friends might think you were weird, Mr. Fletcher probably didn't know anyway, and, after all, it wouldn't be on the exam. So you put off your question again and eventually forgot it, again.

Now you're an adult. You're in your doctor's office for your annual checkup. No particular problems, but out of the blue you remember that question you first wondered about when you were a kid. Should you ask the doctor? After all, doctors are trained in this sort of thing. They ought to know everything about the body because it's their job to fix it when it's broken. But you hesitate. The doctor is busy. There are other patients waiting. And after all, your question doesn't relate to your health or to any illness you're likely to get. So you put off the question yet again.

Has this ever happened to you?

If so, then this book is for you. You can stop putting off your questions about the human body. Chances are the answer is here. *The Odd Body* tries to explain all those body mysteries, both major and minor. We call these OBQs—Odd Body Questions. We ourselves have been asking these sorts of questions for many years—more than we'd like to admit. We love the commonplace questions, the silly, the weird, the bizarre, the fascinating. We hope your question is answered here. Perhaps too there are a few questions within these pages that you never thought to ask. It might be fun to know the answers just the same.

If there's any real lesson in this book, it's simply this: Human beings are *so* interesting. Finding out about us is one of the true pleasures of life.

Chapter 1
Beginnings

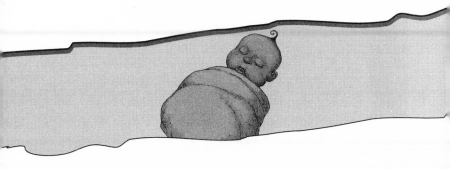

Many of us have questions about our origins, our development in the womb, and how we are born into this world. It's said that we come into this world with nothing, but that's only the beginning of the story.

What makes me a human being?

We are humans because we are classified as such based on our unique physical and cultural characteristics. We manipulate symbols, express ourselves through language, and possess an enormous capacity to develop the intricacies of culture.

Taxonomy is the science of classifying life-forms. As science classifies humans, we are members of the animal kingdom, the metazoan subkingdom, the chordata phylum, the vertebrata subphylum, the class mammalia, the subclass theria, the infraclass eutheria, and

THERE ARE TEN human body parts that are only three letters long: eye, hip, arm, leg, ear, toe, jaw, rib, lip, and gum.

the primate order. After this, it starts to get extremely interesting.

The suborder called anthropoidea is within the primate order. This suborder includes monkeys and apes as well as humans. Within the anthropoidea is the superfamily called hominoidea. This superfamily includes the anthropoid apes and both extinct and modern humans. It excludes the nonanthropoid apes. Anthropoid apes are tailless and include the gibbon, the chimpanzee, the gorilla, and the orangutan. Within the hominoids is the family called hominidea or hominids. Hominids include both modern and extinct forms of human beings. It excludes the anthropoid apes.

What makes the hominids so special is their large brain and their ability to walk on two legs (bipedalism). Where to draw the line between human and humanlike ancestors is a tough decision. One way to do that is simply to include all hominids as humans.

As for the beginnings of the earliest hominids—the beginnings of us—anthropologists have been pushing the date back for most of the last one hundred years as new fossil evidence is revealed.

In 1974, a female hominid skeleton nearly 40 percent complete was found by Dr. Donald Johanson and Tom Gray of the Institute of Human Origins in Berkeley at a site near Hadar in Ethiopia. Nicknamed "Lucy," she was estimated to have lived for forty years and to have attained the height of nearly forty-two inches. Lucy was dated at 3.2 million years old.

In 1978, fossilized footprints and parallel tracks left in volcanic ash and extending over a distance of about twenty-six yards were discovered by Dr. Mary Leakey and Paul Abell near Laetoli in Tanzania. The three obviously hominid beings that left the prints and tracks were estimated to be about forty-seven inches tall. The fossils were dated at 3.6 million years old.

In 1984, a hominid jawbone with two molars nearly two inches long was found by Kiptalam Chepboi in the Lake Baringo region of Kenya. It has been dated at four million years old.

In 1994, Drs. Johanson and William Kimbel, along with Dr. Yoel Rak of the University of Tel Aviv, reported finding fragments of a hominid skull as well as a number of limbs and jawbones at Hadar. These were dated as being about the same age as Lucy, but this hominid was much taller.[1]

Later in 1994, Drs. Tim White, from the Department of Anthropology at the University of California at Berkeley, Gen Suwa from the University of Tokyo, and Berhane Asfaw from the Ethiopian government reported finding part of a child's jaw and two teeth at a site near the village of Aramis, about forty miles south of Hadar. These fossils, together known as Ardipithecus ramidus, have been dated at 4.4 million years old.[2]

In 2001, Drs. Martin Pickford and Brigitte Senut of the National Museum of Natural History in Paris discovered the remains of a creature they called Orrorin tugenensis in the Tungen Hills of Kenya. The nineteen specimens now in this collection, including bits of jaw, arms, fingers, and leg bones (femurs), along with teeth, have been dated at six million years old.

Finally, in 2002, Dr. Michel Bruner from the University of Poitiers in France unearthed in the sandstorm-scoured northern Djurab Desert of Chad a remarkably complete skull. Scientifically named Sahelanthropus tchadensis, it was nicknamed "Toumai" or "hope of life" in the local Goran language. This skull has been dated at seven million years old.[3]

Humans are also Homo sapiens: We belong to the genus of Homo and the species of sapiens.

The earliest member of the Homo genus is the Homo habilus, or "handy man." In 1964, part of a "handy man's" skull was found at Olduvai Gorge in Tanzania and named by Drs. Louis Leakey, Philip Tobias, and John Napier, with the assistance of Raymond Dart. The following year another skull fragment was found in western Kenya but not dated until 1991. The oldest "handy man" remains have been dated at 2.4 million years old.

Homo erectus is the nearest direct ancestor of Homo sapiens. In 1985, Kamoya Kimeu found the earliest remains of Homo erectus at a site near Lake Turkana in Kenya. It was a nearly complete skeleton of a twelve-year-old boy who stood almost five and a half feet tall. The skeleton was dated at 1.6 million years old.

The earliest human tools were found in 1976 by Drs. Helene Roche and John Wall near Hadar. These basic stone implements used for chopping and slicing have been dated at 2.7 million years old.

When did I first know I was alive?

We probably know that we are alive sometime before we are born, but it is difficult to remember this. It is theorized that we fail to remember because we do not have language to hold on to the memory.

The fetus becomes conscious sometime during the second trimester of pregnancy. Tactile sensitivity begins as early as the seventh week, when a fetus first reacts to the stroke of a hair on its cheek. Skin sensitivity expands to include most parts of the body by the seventeenth week.[4] From the sixteenth week on, the unborn baby is easily startled by loud noises and turns away when a bright light is flashed on its mother's abdomen. The fetus reacts actively to heavy-metal rock music by kicking frantically and reacts in the opposite manner to calm music. It is unlikely that anything other than the physical sensation of sound can be detected by a fetus. It is rather like the noise one hears from a distant house where a stereo is blasting away: You can hear the pulsation of the bass but cannot distinguish the lyrics.

Even as early as twelve weeks the fetus can be observed apparently squinting and scowling. At fourteen weeks it seems to sneer and look dissatisfied. But at twenty-four weeks the fetus shows behavior that may indicate true thinking (cognition): frowning, grimacing, and smiling. And while being viewed via ultrasound a fetus at twenty-four weeks who was accidentally hit by a needle during an amniocentesis was observed twisting its body away, locating the needle with its arm, and repeatedly striking the barrel of the needle with its arm and hand.[5]

There is speculation that the fetus must be thinking when it demonstrates anxiety. At twenty-four weeks the fetus seems to be

ACCORDING TO anthropologists, life in the Stone Age might not have been too much fun. Our ancestors wore the skulls of their dead ancestors as mementos, ate elephant meat raw, smeared their bodies with animal grease to keep warm during the coldest winters, washed their bodies with dirt, and kept wolves as pets.

anxious, because it can be observed sucking its thumb—sometimes so hard that blisters are raised.

By twenty-six weeks the fetus can do some rather interesting gymnastics inside the womb—for example, an elegant forward roll. Whether such movements are intentional and thus indicative of thinking is a matter for speculation.[6]

When did I first dream?

Other evidence of thinking has to do with dreaming. There is evidence that the fetus dreams. Indeed, the fetus dreams more than newborn infants, who in turn dream more than older children, who dream more than adults. Sonographic studies show that rapid eye movement sleep (REM sleep, in which dreaming takes place) occurs at twenty-three weeks. This is virtually the only type of sleep the fetus engages in. Non-REM sleep is not detected until thirty-six weeks, so one could almost say that after twenty-three weeks whenever a fetus is sleeping it is dreaming.[7]

When did I first feel?

Solid evidence suggests that the fetus feels pain by no later than twenty-six weeks, yet some claim this ability occurs much earlier. One study suggested that the fetus feels pain as early as the seventh week.[8]

Pain pathways in the brain, as well as the cortical and subcortical centers necessary for pain perception, are well developed by the third trimester. Responses to painful stimuli have been documented in newborns (neonates) of all viable gestational ages.

In 1969, Dr. Davenport Hooker at the University of Pittsburgh found that a fetus aborted during the thirteenth week (but not yet dead) will respond reflexively to the touch of a hair around its mouth. He also reported that a baby born three months prematurely will respond reflexively to the touch of a hair anywhere on its body.[9]

There is every indication that a newborn baby is in some ways just as sensitive to touch as older humans are. A newborn's skin is thinner than an adult's, so its nerve endings are less well insulated. Moreover, a newborn's nerve endings are just as mature and far more numerous than an adult's. The portion of the brain that processes touch sensations (the somatosensory cortex) is at birth more developed than any other portion of the brain.[10]

IF STONE AGE people were not feeling well, they sometimes cut holes in each other with a sharp stone to let the pain escape.

BETWEEN fertilization and birth, a baby's weight multiplies an extraordinary five million times.

Nevertheless, it takes years for the sense of touch to fully develop. Children cannot distinguish most objects by touch alone until they are about six or seven years old. The first fetal touch receptors appear on the skin by no later than the tenth week—while still surrounded by water. However, according to Dr. Maria Fitzgerald, professor of developmental neurobiology at the University of London, "although the fetus lives in fluid, it never feels wetness."[11] It is just like a person

A TERRATOMA (sometimes spelled without the second "r") is an unusual tumor, cyst, or other growth produced by the body. It is usually benign, but not always. The word literally means "grows out of your body."

swimming under water who will not feel the water as such but "will notice the pressure of the wave."[12]

When did I first see?

Vision develops to some extent before birth, but the newborn is very nearsighted. The fetal eyelids form by ten weeks but remain fused shut until the twenty-sixth week at the latest. Nevertheless, the fetus will react to lights flashed on the mother's abdomen.[13]

Visually, babies are fascinated by two things: the human face and high-contrast geometrics. The general thrust of research in this area leads to the following conclusions:

From birth to about two months of age, babies see objects best at close range: about 11.5 inches from the eyes at birth and about 11.8 inches away at six weeks of age. Babies can discriminate between differences in shape, size, and pattern and are more attracted to high-contrast patterns than to color or brightness alone. They prefer to look at patterns of simple to moderate complexity, and they look more frequently at outside edges than internal patterns.

From about two to four months of age, babies scan their entire vision field and explore both interior patterns and outside edges. They prefer patterns of increasing complexity and round curved lines and shapes to straight lines or angular shapes. They are especially attracted to faces and shapes of all kinds. Babies begin to show that they remember what they see.

After about four months of age, babies adjust their focus to see near or far objects. They see in full color and continue to prefer curved patterns and shapes. They seek out complexity and novelty in their visual environment and begin to develop depth perception.[14]

Children usually learn to identify colors between the ages of three and seven. If they seriously confuse colors after this time, color blindness is a distinct possibility.

When did I first hear?

The fetal listening system begins to function by sixteen weeks—even before the ear is complete.[15]

Surprisingly, the sense of hearing in a fetus begins with the skin. According to Dr. David Chamberlain, author of *The Mind of Your Newborn Baby,*[16] the skin is "a multisensory receptor organ integrating input from mechanoreceptors, thermoreceptors, and pain receptors (nocireceptors). This early form of hearing is linked with the vestibular system, which is sensitive to gravity and space, and with the cochlear system as it forms."[17]

The hearing of the newborn is excellent. We have known this for decades, from experiments with the startle reflex. In fact, hearing in the newborn is far more mature than vision. A series of classic experiments demonstrated that before the infant is fully delivered from the birth canal, when just the head is popping through, the infant's eyes will turn toward a sound made at one side or the other of the head. It's as if the baby knows that something is there to be seen.[18] It's also interesting to note that the newborn hears as well when sleeping as when awake.

When did I first smell?

Some people always smell! It depends on how you mean this question. But seriously, the fetus can definitely smell, even though it is surrounded by fluid. However, according to Dr. Stephen Roper, professor of anatomy and neurobiology at Colorado State University in Fort Collins, "the fetus doesn't sniff. Rather, odors are absorbed by nasal tissues."[19] Indeed, many species of fish have the same ability.

The amniotic fluid in which the fetus swims is full of odors. If the mother eats spicy food, the fluid can smell like a Mediterranean salad. Also, the subtle smells in the fluid are unique to the mother—just as body odor is unique. After birth, these smells may help solidify the baby-mother relationship.

Immediately after birth, the newborn cannot smell through its nose very well because the nose is clogged with amniotic fluid and other substances for a day or so. This clogging effect resembles the stuffiness of an adult's nose during a cold. A baby's sense of smell starts to emerge right after the clogging stops—two days after birth. There is even evidence that babies who are just a few days old have as good a sense of smell as adults.

As early as 1934, Dr. Dorothy Disher found that one-month-old babies were more likely to wriggle in their cribs when presented with a

A NEWBORN baby expels the equivalent of its own body weight in poo every sixty hours.

HUMAN BIRTH-control pills work on gorillas.

HUMAN babies are less intelligent at birth than chimpanzee babies.

number of odors, compared with smelling merely pure air. Babies responded most to the smells of violet, asafoetida, sassafras, citronella, turpentine, pyridine, and lemon.[20]

In a now-classic laboratory experiment, Dr. Jacob Steiner of the Hadassah School of Dentistry at the Hebrew University in Jerusalem asked a panel of adults to select the most "fresh" and the most "rotten" from among a large collection of odors. The adults judged that honey was the freshest, followed by banana, vanilla, and chocolate smells. Rotting eggs followed by rotting prawns were unanimously judged the most rotten smells. Dr. Steiner then held swabs of these odors under the noses of babies only a few hours old. They smiled when smelling the fresh odors and grimaced when smelling the rotten ones. Furthermore, the widest smiles came from smelling the honey and the biggest grimaces came from smelling the rotting eggs—identical to the choices made by the adults.[21]

Other researchers have found evidence that a majority of newborn babies can smell *better* than adults—in this case, the researchers. In 1975, Oxford University psychologist Dr. Aidan Macfarlane tested whether a newborn could tell the difference between the smell of its own mother (and her milk) and the smell of another baby's mother (and her milk). The smells came from gauze pads that the mothers had kept within their bras to absorb any milk leaking from their breasts. Dr. Macfarlane draped a pad from the baby's mother along one side of the baby's face. Along the other side, he draped a pad from another mother. More than two-thirds of the six-day-olds tested "turned toward their mother's pad, as did more than three-quarters of the eight- to ten-day-olds. Young babies prefer the familiar to the unfamiliar. Here they recognized their mother's odor, and turned toward it. Although babies less than six days old did not turn toward their mother's pad, the older babies certainly did smell a difference—a difference that Macfarlane, when he smelled the pads himself, was unable to detect. . . . Macfarlane's babies did not just detect an odor, they *recognized* one. To recognize something requires high-level processing within the brain—some kind of conscious processing beyond the reflexlike processing of the midbrain. . . . Macfarlane's study does make it look as though the newborn's sensitivity to smells is close to being adult."[22, 23]

When did I first taste?

The fetus can taste at fourteen weeks, which is when all the tasting mechanisms are in place. Swallowing can be seen via ultrasound. By the end of the first trimester, the fetus controls the frequency of its own swallowing in response to sweet or bitter tastes.[24] The fetus regularly swallows amniotic fluid. Thus, according to Dr. Gary Beauchamp, director of the Monell Chemical Senses Center in Philadelphia, a fetus swims in a "smorgasbord" of flavors—the sweetness of glucose, the saltiness of sodium, the bitterness of its own urine. According to Dr. Tiffany Field, director of the Touch Research Institute at the University of Miami School of Medicine, "amniotic fluid is kind of brackish-tasting, and we have videos of a fetus grimacing when it swallows the fluid."[25]

For nearly seventy years we've known that humans have a definite preference for sweet-tasting things over bitter-tasting things.[26] Newborns are very good at distinguishing tastes. They stick their tongues out when they taste sour things. Babies whose mothers frequently ate garlic during pregnancy show a specific preference for garlic-flavored foods. We think that's because they associate the garlic taste with mom—and of course they love mom!

How soon was I able to cry and laugh?

Audible crying of a fetus has been recorded as early as twenty-one weeks.[27] Laughing occurs much later, at about six months of age. Smiling is something else again. Research by Dr. Susan Jones at the Department of Psychology at Indiana University has firmly established that we all have the equipment to smile from birth. Her most recent work involves observations of eighteen-month-old toddlers. She has also demonstrated that babies stop smiling very quickly if no one pays attention to them.[28]

How soon was I right-handed or left-handed?

Research on fetal thumb-sucking behavior shows that handedness starts in the womb. When fetuses are observed by ultrasound imaging, some already indicate a preference for one hand over the other as early as fifteen weeks gestational age. Dr. Peter Hepper led the team of researchers from the Fetal Behavior Investigations Unit at the Queen's University of Belfast, Northern Ireland.[29]

FOR SOME time now, chimpanzees have been taught sign language. Today, some of the original learners have taught the language to other chimpanzees such that they have conversations among themselves.

Was I born with more than
five senses?

Synesthesia—the merging of senses—is a fascinating phenomenon. It may be that we are born with senses that are undifferen-
tiated and that become separated into our recognized five senses of
sight, smell, hearing, taste, and touch only sometime after birth.

We learn in school that each sense is separate and relates to a particular body organ. We hear sounds with our ears and see sights
with our eyes—never vice versa. Indeed, we are led to believe that
our senses are distinct, individual, and completely independent of
the others. Thus, people who are totally blind may still have excellent hearing. We are also taught to assume that a human being cannot see sounds, hear sights, or touch tastes, so that concept is
completely foreign to us.

The process wherein an individual experiences confusion of
senses—as if the five senses were somehow merged or fused into
one—is known as synesthesia. Sounds *are* seen, sights *are* heard,
tastes *are* touched, and so on. A growing body of evidence shows
that synesthesia can occur in adults, although it is extremely rare.
Notable so-called "synesthetes" include the composers Olivier
Messiaen, Aleksandr Scriabin, and Nikolai Rimsky-Korsakov. In
fact, one alleged adult synesthete, the anonymous Russian vaude-
ville memory expert known simply as "S," was studied for almost
thirty years by a Russian psychologist.[30]

Stronger evidence exists that we are all synesthetes as newborns.
Our five senses are at first blurred but become more specialized as
we mature and as our sensory channels to and from the brain
become fully developed.

Dr. Robert Hoffmann of the Department of Psychology at Carleton University in Ottawa attempted to measure the speed at which sensory impulses reached the brain in newborn infants between one and three months old.[31] In an experiment, he attached electrodes to an infant's head, exposed the infant to flashing lights from translucent panels, and then recorded the speed of the electroencephalogram (EEG) waves. Readings were taken from directly over the visual cortex as well as from three areas of the

IF YOU COULD remove all the space from the atoms that make up your body, you would be small enough to pass through the eye of a needle.

brain far removed from the visual cortex. Dr. Hoffman reported:

These waves came through all four EEG electrodes: they welled up from all over the cortex. They were formed by energy from the eyes that was channeled throughout (and amplified by) the brain, where it would impinge upon, and mix with energy flowing through, other neuronal channels. Such impingement is the stuff of thought—not just verbal thought, but all associations both meaningful and confused. . . . This showed that all of the babies did indeed perceive the stimuli directly—but these differences came not from the direct sensations; these differences came from the impingement of varying amounts of visual energy upon non-visual areas of the brain. There energy from the eyes might mix with energy from the ears to produce vague sounds; or it might mix with energy innervating muscles to cause a twitch—which would in turn fire sensory neurons within those muscles, causing the sensation of movement. The amount of mixing depends on the amount of energy entering the nervous system. This total amount of energy added to the nervous system is a prime determinant of a newborn baby's perceptions.[32]

Drs. David Lewkowicz and Gerald Turkewitz from the Albert Einstein College of Medicine in New York discovered that babies between three and four weeks of age "equate brighter lights with louder sounds."[33] In an experiment, the two researchers first asked adults to attempt to adjust the volume of a loudspeaker to make it equal the brilliance of a light. The adults agreed remarkably well on the level of noise that seemed to equal the intensity of the light. Next, twenty infants were repeatedly exposed to the light while their pulses were monitored. A burst of noise was then substituted for one of the flashes. Although the level of noise the adults had decided was equivalent to the light caused little reaction, every other level caused a marked quickening of each infant's pulse. And the quickening was proportionate to the difference between the intensity of the light and the intensity of the matching sound.[34]

These two experiments suggest that the senses are intertwined in infants, that synesthesia exists in the infant's brain.

Can we learn to recapture in adulthood our previous state of infant synesthesia? And could we ever learn to see without eyes or hear without

THE MOST FETUSES found in a human body—fifteen, ten girls and five boys—were four months old when they were removed from the womb of an Italian housewife in July 1971. The woman had been taking a fertility drug.

ears by drawing on our other senses? If so, it may be possible to retrain people who have lost one sense to draw on another.

Dr. Richard Cytowic, a neurologist in Washington, D.C., and author of the two best-known books on adult synesthesia, believes that this just might be possible.[35, 36] He has seen more than forty synesthetes in his practice and is among the foremost authorities on the subject. According to Dr. Cytowic, the sensory world of his patients is one of salty visions, purple odors, square tastes, and green, wavy symphonies. One of his patients even allegedly has technicolor orgasms. Nevertheless, Dr. Cytowic believes that understanding synesthesia will provide the key to understanding the mind.

What is the most common form of synesthesia? That seems to be "color-hearing." According to Dr. Simon Baron-Cohen of the Department of Psychology at the University of London, color-hearing synesthetes nearly always have "colored vowels" and "colored letters."[37] When they hear vowels or read letters, they report "seeing" colors.

How common is synesthesia in adults? Of course it is rare. Approximately one in twenty thousand people are synesthetes. It is interesting that "when he [Dr. Baron-Cohen] reported the [1987] findings in [a radio interview], more than two hundred women (and two men) wrote in, claiming to have synesthesia—an astonishing response, given that it was a science program with at least equal numbers of male and female listeners."[38–40]

Related to synesthesia, but very different, is "blindsight." Some blind people possess a form of "unconscious vision" that allows them to "see," after a fashion, without being aware of it. That is blindsight.

The term "blindsight" was coined in 1980 by Dr. L. Weiskrantz of the Department of Psychology at Oxford University and colleagues at the National Hospital in London. Here is what the researchers discovered:

> *Working with patients who had acquired visual defects as a result of damage to the brain cortex, they found that their subjects, if asked what they saw, said they saw nothing. In testing, though, asked to guess, they "guessed" right nearly 90 percent of the time. This is far beyond the success rate accountable to pure chance. The researchers theorize that the kind of "sight" detected by these tests depends on a different neural pathway from the eye to the brain, one that passes through the midbrain rather than the cortex. In evolutionary terms, the midbrain is much older than the cortex. In a sense, it may be the animal part of our brain—an evolutionary holdover from our prehuman past.*[41]

Other experiments with blind people who have blindsight found that some can catch a ball tossed to them, "even while insisting they can't see anything."[42] In any case, deprived of vision, people with blindsight somehow do seem to draw on other senses to "see." Scientists have so far been unsuccessful in explaining blindsight abilities.

How early can I conceive a baby?

Girls can begin conceiving at puberty. The average age of puberty has decreased by about two and a half months with each generation since the end of the nineteenth century. A Brazilian girl reputedly gave birth at six years, seven months, and three days of age. The oldest mother, from Oregon, gave birth at fifty-seven years, six months, fifteen days of age, and without the use of fertility drugs. However, modern drugs and artificial fertilization techniques have pushed this age later and later. Theoretically, there is no upper age limit on when we can no longer conceive children, but who wants to chase after a toddler at age seventy?

> MORE IN-VITRO babies are born in Australia than in any other country.

How many babies can I have?

Before the era of fertility drugs, an eighteenth-century Russian woman gave birth to sixty-nine children, of whom sixty-seven survived into adulthood. She was able to do this by giving birth to sixteen pairs of twins, seven sets of triplets, and four sets of quadruplets along the way.

Every human female possesses some two million eggs (ova) at birth. Of these, about three hundred thousand survive to puberty. And of those only 450 are ultimately released for possible fertilization—one each month during the reproductive years (roughly from twelve to fifty years of age). The human male produces half a billion sperm each day. Four hundred million are released in a single ejaculation. Men can remain fertile somewhat longer than women.

Assuming that a monogamous couple has sexual intercourse often enough that all sperm produced are released, and assuming that a man remains fertile for fifty years, the chance of any one sperm fertilizing any one ovum is 18,263,000,000,000,000,000 to one.

Is "immaculate conception" possible?

Reproduction without sperm is called parthenogenesis. This can occur in some plants and in invertebrates. It can also occur in some species of insects, fish, reptiles, amphibians, and birds. Honeybees, wasps, and some lizards are examples. But it does not occur in mammals—including humans.

Experiments at Yale University have attempted to induce parthenogenetic development in mice. In these experiments, ova start to develop after exposure to three factors: electric shock, mechanical agitation, and a saline solution. However, the embryo always dies before the halfway point of gestation.

How much did I weigh at birth?

Ask your parents about this one. Nine out of ten babies are born weighing between 5.29 pounds and 10.58 pounds. Male babies are slightly heavier than girls on average (the difference is less than half a pound). The heaviest baby on record weighed twenty-nine pounds at birth. The lightest baby to survive weighed not quite ten ounces. For some unknown reason, babies born in November in the Southern Hemisphere, and in May in the Northern Hemisphere, weigh on average six ounces more than babies born in any other month.

It is also interesting that more babies are born between midnight and 8:00 A.M. than during the two other eight-hour periods.

Tuesday is the most popular birth day, while Sunday is the least popular. More babies are born during days with a full moon than during any other time in the lunar cycle.

For unknown reasons, babies born to brown-haired mothers are delivered slightly faster than babies born to blonde-haired mothers.

Are babies born without scars?

At birth, many newborns look like they've boxed twelve rounds with Mike Tyson. Babies delivered with the help of forceps tend to be somewhat bruised but heal very soon afterward. However, in utero surgical procedures, which are now undertaken around the world, have resulted in the discovery that the fetus does not scar.

Why are babies born without teeth?

Ask any breast-feeding mother why nature builds this one in! Less pain for the nursing mother means the baby is less likely to be rejected. Strangely, approximately one in every two thousand babies has at least one erupted tooth at birth. Even more strange is that a large number of famous world leaders, including several emperors and dictators, are known to have already cut a tooth at birth. Julius Caesar, Hannibal, Charlemagne, Napoleon, Mussolini, and Hitler are examples. Could it be that the mother's pain during breast-feeding caused the mother to react negatively, withhold love, and even reject the child emotionally if not physically? In turn, did this rejection and denial cause the child to seek world power and domination later, as a substitute? It's fun to speculate. What would Sigmund Freud say?

Will we ever be able to make an android to replace the human body?

We are near the time when we can replace human beings with robots in human form. The specter of Arnold Schwarzenegger as "The Terminator" or Lt. Commander Data of *Star Trek: The Next Generation* is becoming less of a science-fiction fantasy and more of a science fact.

According to one scientist, "the age of intelligent androids—not merely better computers, but a new form of life—may be as little as twenty years away." Dr. Maureen Caudill, a researcher at the Artificial Intelligence Laboratory at the Massachusetts Institute of Technology, has surveyed the achievements of various technologies required to make androids a reality. It is a new and fascinating research field in which advances occur almost monthly.[43]

Already simple robots can navigate around a room or pick up an egg without breaking it—a task even experts once thought impossible. Far beyond this, there are now artificial neural networks closely approximating the human brain, along with sophisticated vision systems, memory systems, language systems, pattern recognition systems, and other processes. All these processes would have to be replicated to a human level by any self-respecting android.

A robot may soon be carrying your baggage through the airport.[44] Autonomous mobile robots equipped with sensing devices and artificial intelligence will soon perform a number of similar duties. Some potential robot jobs include house cleaners, office-mail carriers, minefield sweepers, and underwater or outer space–based construction workers.

Robots will soon be used in the travel industry to greet guests (in any language), babysit, and perform security tasks.[45] The night watchman may soon be a thing of the past. Admittedly, robots may not have "the human touch," but they will work tirelessly and not complain when they are put in a closet during the off-season.

A new "artificial muscle" is being made for robots. The muscle, made from a gelatin-like substance, has been developed by Dr. David Brock at the Artificial Intelligence Laboratory at M.I.T. This invention may lead to smaller, stronger, and more flexible robots—bringing us another step closer to the age of androids.

More specifically, the Brock artificial muscle consists of polymer gel fibers. Polymer gels change their volume quickly in response to changes in pH. Thus, adding a basic solution makes it expand. By alternating of acid and basic solutions, the muscle can be made to lift or lower a three-and-a-half-ounce weight.

HALF THE world's population today is under twenty-five years of age.

FOR ROUGHLY six to seven months after birth, an infant can breathe and swallow at the same time. Older children and adults cannot do this.

This artificial muscle may someday replace many of the gears, pulleys, and motors that today's robots require. In addition, robots could be free from dependence on electrical outlets because the muscle is operated solely by changes in pH.[46]

Dr. Caudill describes a Japanese robot, the WABOT, that can read simple sheet music, play keyboard instruments, and accompany a human singer. And a machine-vision system developed by German researchers has successfully guided a car without human control at speeds up to nearly sixty miles an hour. Dr. Caudill also notes progress in speech generation and speech recognition systems (versions of which are already being installed in some computers).

The key technology for building androids will be sophisticated neural networks. Dr. Caudill explains that "neural networks are information processing systems physically structured in a way that mimics our current understanding of the brain."

Neural networks differ from digital computer systems. Instead of following the instructions of a program or pigeonholing every bit of data in a specific location, as a digital computer does today, neural networks—like nerve cells that have many connections with one another—function by stimulation through patterns of activity. Dr. Caudill argues that "these patterns can be *trained*, rather than programmed, to achieve desired results." Furthermore, "neural networks also differ [from older technology] in other ways. Older artificial intelligence systems emphasize developing intelligent behavior through logical step-by-step procedures, while neural networks are much less structured and much more flexible."

Dr. Caudill notes that neural network research is providing models for testing how human brains work, learn, and organize themselves—the fundamentals of human psychology. One experimental neural network, using a new psychological model, has successfully imitated the classical conditioning process by which animals are trained. This process was first described by Ivan Pavlov more than a century ago and represents a cornerstone of human psychology. Another new

ALL THE PEOPLE on Earth together weigh about 420 million tons.

FOR EVERY person on Earth, there are two hundred million insects.

neural network is modeling the internal "mapping" that occurs when a newborn baby's brain absorbs its first environmental stimuli outside of its mother.

Thus, by trying to build an artificial brain—and ushering in the android age—we are also learning more about ourselves.[47]

Can we ever live forever?

We humans may very well and very soon come quite close to achieving immortality. Aging may become a thing of the past. And if aging is licked, then all the age-related diseases that kill us will also be licked. That would bring us close to living forever.

As impossible as it sounds, genetically engineered drugs that reverse the aging process will be available in the near future, making it possible not only to halt the aging process but also to turn the clock back. At age seventy we will be able to look and feel as we did at fifty, and better than we did at sixty. And should the age reversal process continue for another ten years, at eighty we could look and feel the way we did at forty.

Age reversal is possibly based on what scientists already know about the process of aging itself and about drugs to cure diseases. That same technology is being applied to aging.

Science tells us that we age because of problems with individual cells in our bodies. As it were, the whole is equal to the sum of its parts. Cells are continually being replaced by other cells throughout our lifetime, but the cells produced later in life are known to have more defects, mutations, and so-called DNA "spelling errors" in them than the cells produced early in life. As time goes on, there are more of these error-prone cells than there are cells that are error-free.

IF YOU GAVE EACH human on Earth an equal portion of dry land (including the uninhabitable areas), everyone would get roughly thirty-three square yards.

Our bodies do not look as good, and we notice they do not function as well either.

Why do our cells commit more "spelling errors" when we get old? This is probably because of the cumulative damage from free radicals within the cell. Free radicals are molecular fragments produced during normal cell metabolism that react wildly and unpredictably. Radiation, toxins, carcinogens, stress, and other factors can aid the production of free radicals. Furthermore, although the body produces enzymes that serve as a kind of "correction kit" for the DNA "spelling errors," these enzymes are not produced in as great a supply as we get old. As a result, the older we are, the more likely our replacement cells will have errors that are "uncorrected"— just like a spelling error in a word.

Based on this theory of the cell replacement process, scientists are working to devise ways to boost and extend the life of the DNA-repair system. This would mean both correcting the "spelling errors" that can occur when DNA replicates itself when each new cell is born, and repairing DNA damage caused by free radicals. The reversal of aging would occur because error-prone, damaged, "old" cells would eventually be replaced by error-free, undamaged, "new" cells, until the entire body is composed of only youthful cells.

Although it may seem a mere dream, this is no science-fiction fantasy. It is theory being turned into reality in laboratories around the world. At the Center for Molecular Science at the University of Texas Medical Branch in Galveston, Dr. Samuel Wilson is undertaking experiments leading to animal experiments and eventually to human experiments that will result in drugs that humans can take to transform old age into youth.

Dr. Wilson has already "isolated the gene in mice for one of DNA's key repair enzymes in both humans and mice. The production of this enzyme is known to decline steeply with advancing age. His plan is to produce a line of mice that carry many extra copies of the gene for this critical enzyme, in the hope that those extra genes would keep turning out an abundance of the enzyme. These would be on hand to carry out the genetic repair work for a much longer period of time. Thus, the animal's DNA would accumulate errors and mutations at a much slower rate, and the mice would live to a venerable age." Dr. Wilson said in 1992, "I would guess it will be a year or more . . . before we can report that we have a successfully engineered mouse. After that, it will only take another six months before we can tell for sure whether we have increased their life span."[48] By 1998, genetic engineering experiments with mice resulted in the doubling of their normal life expectancy. By 2001, it tripled.

Another team at Galveston is headed by Dr. John Papaconstantinou. Research by Dr. Papaconstantinou is still in its early stages. His group is working with several genes, including those involved in the response to everyday stress. It is interesting that the proteins these genes produce are able to maneuver themselves from the cytoplasm of the cell into the nucleus, where they can search out specific sites on the DNA and turn selected genes on and off. According to Dr. Papaconstantinou, "as we age some cells make too much of some proteins and not enough of others." This means that genes are being turned on or off inappropriately. Thus, "our long-range hope is to learn enough about these processes to enable us to bring the cell back to its young-adult balance by manipulating the expression of the appropriate genes."[49]

If Drs. Wilson, Papaconstantinou, and others around the world are correct in theory and successful in practice, their work could open the door to gene treatments promising age reversal—changing our cells, and hence ourselves.[50] When will all this happen? The best estimate is that by 2010 at least some longevity drugs will be available for humans.[51] According to Dr. Michael Fossel of Michigan State University, at the latest, living "a couple of centuries" is only "a couple of decades away."

In any case, Dr. Thomas Perls of Harvard Medical School notes: "Traditional views of aging may need rethinking." People are living longer now and are healthier than ever before. And "people in their late nineties or older are often healthier and more robust than those twenty years younger."[52]

So there's every chance, more so than ever before and one way or another, that many of us may hit the century mark—and beyond.

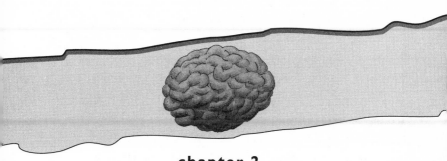

the **brain**

The United Nations designated the 1990s as "The Decade of the Brain." We learned much about the brain's many wonders in the 1990s, but there is still much more to learn about what Woody Allen once called his "second favorite organ."

What is the brain anyway?

This is going to be a long explanation, but here goes. The human body functions as an interdependent, coordinated unit. Its activities, actions, and reactions are directed by the brain through the vast and complex network of the nervous system. The brain itself is the most complex and largest mass of nervous tissue present in the body. It directs the complex activities of the body. Through the five senses—touch, sight, smell, taste, and hearing—the brain keeps each of us in touch with conditions and events in the world around us.

The human body performs two basic types of movements or actions: voluntary and involuntary. In a voluntary action, the brain calls on the muscles or organs of the body to perform a task. In an involuntary action (a reflex), the senses communicate a condition or situation to the brain (a stimulus), and the brain responds by calling on motor nerves to react or respond to the stimulus. In a reflex, the brain is often bypassed. In such a case, the sensory nerve may call on the motor nerve indirectly via the spinal cord for the required response, or directly via contact with the motor nerve. A reflex does not require thought or "brain work" as such in order for the proper action-reaction response to take place.

Most of the nerves that activate the muscles of the body stem from the spinal cord. The spinal cord runs up the spine and enters the skull cavity through an opening at the base of the skull (cranium). It then expands into the medulla and then to the cerebellum to form the base of the brain. Sitting on the cerebellum are the brain's remaining major, distinct, but interconnected parts, which together fill the skull cavity: the pons, the midbrain, and especially the cerebrum.

The weight of the brain of the adult male is approximately three pounds, and the adult female brain weighs slightly less, about two and three-quarters pounds. There is no significance in the gender difference. Males generally have larger brains because they generally have larger bodies.

After conception, the brain of a human fetus appears to be electrically silent for about the first six weeks of life. After this time, intermittent "slow wave" activity of low intensity occurs. In the human embryo, the brain first consists of three sections: the fore-brain, the midbrain, and the hindbrain. As the embryo develops, the brain forms and develops its other parts. Brain growth continues very rapidly up to the fifth year, then continues more slowly until about the twentieth year. It remains stable in size throughout middle adulthood, and then in advanced age it gradually loses weight.

The ability of humans to remember and utilize past experience, to cope with current situations, to think and reason, and to conceive of new, never-before-thought-of thoughts clearly differentiate humans from all other animals.

The cerebrum is the largest section of the brain and the center of intelligence, sensation, emotions, and volition memory. It is divided into two sections—the left cerebral hemisphere and the right cere-bral hemisphere—by a long cavity called the longitudinal fissure. Each hemisphere is made up of an outer coating, the cerebral cortex (the so-called "gray matter"), covering an inner mass of white mat-ter. But within each hemisphere is a space or ventricle connecting with the all-important third ventricle through an opening called the foramen of Munro. These spaces within the hemispheres are collec-tively called the lateral ventricles. The lateral ventricles are spaces within the cerebral hemispheres of the forebrain that are also filled with cerebrospinal fluid.

Each hemisphere has five lobes: the frontal lobe, the parietal lobe, the temporal lobe, the occipital lobe, and the insula lobe. The meninges are the three coverings that lie between the skull and the brain. Moving from the skull downward:

 The dura mater is the tough, fibrous membrane—rough on the outer surface but somewhat smooth on the inner surface—that protects the brain. It contains arteries, veins, and sensory nerves, and projects into the cranial cavity, which divides the cavity into partitions: the falx crebri, separating the cerebral hemispheres; the falx cerebelli, partially separating the cerebellar hemisphere; and the tentorium cerebelli, separating the cerebrum and the cerebellus.

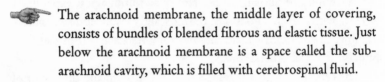 The arachnoid membrane, the middle layer of covering, consists of bundles of blended fibrous and elastic tissue. Just below the arachnoid membrane is a space called the subarachnoid cavity, which is filled with cerebrospinal fluid.

The pia mater is the innermost protective layer of the meninges. It consists of small arteries, veins, and connective tissue that help furnish the blood supply of the brain. The pia mater actually dips into the convolutions of the brain.

Long before birth, as the brain develops from embryo to fetus, the walls of the forebrain thicken and the cavity spaces shrink to form the thalami. Eventually, the cavity spaces of the lateral ventricles are reduced in size to a tiny slit known as the third ventricle.

The midbrain, as the central nervous system develops, undergoes a thickening of its walls, which develop into two cylindrical bodies (the cerebral peduncles) and its central cavity, which is also reduced to a narrow canal.

The hindbrain includes the pons in the frontal section above the medulla, or bulb, in front of the cerebellum. The pons is made up of a bridge of fibers that connect the halves of the cerebellum, joining the midbrain with the medulla below it. The medulla lies between the pons and the spinal cord and contains such vital centers as the respiratory, vasomotor, and cardiac centers. In the hindbrain is a large cavity, the fourth ventricle, that connects with the cerebral aqueduct above and with the central canal of the spinal cord below.

The spinal cord contains nerve cells along its entire length and has bundles of long nerve processes, or nerve trunks, extending to different parts of the medulla, the pons, the midbrain, and the cerebrum.

The cerebellum, the largest part of the hindbrain, lies in the posterior cranial fossa and is covered by a layer of dura mater called tentorium that also serves to separate it from the posterior section of the cerebrum. It is composed of two hemispheres with a middle section or lobe between them called the vermis. Bands of fibers, peduncles, connect the brain stem to the cerebellum. The superior cerebellum peduncle connects the medulla. The cerebellum functions as a reflex center for coordination and degree of voluntary movements. Damage to the cerebellum affects coordination. Yet as reeducation of the motor area occurs, we have learned, voluntary movements are coordinated in the cerebellum but do not originate there.

The cortical areas control particular motor, sensory, and association responses:

 The postcentral area controls sensory activity, touch, and muscle. The visual center is in the occipital lobe; the auditory

MORE ELECTRICAL impulses are generated in one day by a single human brain than by all the telephones in the world.

THE WEIGHT OF your body is forty times greater than the weight of your brain.

center is in the superior temporal convolution; and the taste and smell centers are in the hippocampal area.

 The frontal area controls association of ideas, conduct, behavior, and intellectual concentration.

 The precentral areas control voluntary movement and exercise volitional control over the skeletal muscles. In the front of this area is a psychomotor area that is is involved with carrying out skilled acts.

The basal ganglia (or cerebral nuclei) are gray masses deep within the white matter of the cerebral hemispheres. The most important are the thalamus and the corpus striatum. The thalamus is oval-shaped; in two parts, which are separated by the third ventricle area; and connected with a stretch of gray matter termed the massa intermedia. It appears that the thalamus functions as a center for crude and uncritical sensations and responses. In most animals, including humans, it is the stongest and most sensitive response area.

In humans, the thalamus passes fresh relays of nerve fibers to the cerebral cortex, where finer interpretations and reactions enter conscious sensation. The function of the corpus striatum has not yet been clearly defined, but it seems to have an effect on steadying voluntary movement without initiating such movement.

The hypothalamus is found below the thalamus and forms the bottom and a portion of one

THE HUMAN brain makes up about 2.3 percent of the average person's body weight. By contrast, the squirrel monkey's brain makes up about 5 percent of its body weight. This is the largest brain-to-body weight percentage of any other animal.

wall of the third ventricle. It holds the temperature control centers of the body. A center for controlling heat loss functions through sweating and panting is found in the anterior (frontal) section. Another center for preventing heat loss and increasing heat production is found in the posterior (rear) section. It also plays a part in the metabolic process through a connection with the posterior lobe of the pituitary gland.

Can we live normally with only half a brain?

Some people *have* to live with only half a brain, for their own health.

Serious cases of Sturge-Weber Syndrome involve major brain and body problems—for example, port wine stain birthmarks on the face, especially around the eyes and forehead; pressure on the eyes, resulting in serious glaucoma that can eventually leave a patient blind; epileptic-type seizures that can be very frequent; a lack of body coordination on one side of the body; learning disabilities; and mental retardation.

According to Dr. Steve Roach, a neurologist in Denver and a consultant to the Sturge-Weber Foundation of Aurora, Colorado, when medications are ineffective in arresting the seizures "a [surgical] procedure to remove the hemisphere" causing the seizures has been available for several years. This procedure is called a hemispherectomy. Half the brain is surgically removed. Dr. Roach adds that there is "surprisingly little neurological impairment after the procedure."

Dr. Roach notes that a less radical procedure, but one less likely to work in stopping the seizures, is called a corpus callostomy. The hemispheres of the brain are surgically divided, but neither is removed.[1]

Why are most people right-handed?

Perhaps it is odd that a chapter on the brain would discuss a matter of the hands, but such is the unusual character of this book. You see, your brain determines your handedness.

The left side of the body is controlled by the right hemisphere of the brain, and the right side of the body is controlled by the left hemisphere of the brain. Right-handers are left-hemisphere dominant while left-handers are right-hemisphere dominant.

About 88 percent of humans are right-handed. That leaves about 11 percent left-handed. However, it depends on how handedness is defined. Some people favor one hand for doing some things and the other hand for other things. A truly ambidextrous person—one who is equally hemisphere dominant and who equally uses either hand—is quite rare.

According to science writer Marc McCutcheon, "most cases of left-handedness . . . are thought to be caused by minor brain damage before or during birth. Many scientists believe the damage is due to reduced oxygen supply before birth."[2]

Twins have a higher than usual rate of left-handedness. This is believed to be caused by twins having to share less space in utero and perhaps receiving less oxygen.

Sixty-five percent of those suffering from autism are left-handed. Left-handedness is found to be more common than usual among the world's artists and also among the world's gays.[3]

Are identical twins either both
right-handed or both left-handed?

Although identical twins are genetically the same, all sorts of things about their behavior differs. Handedness is one of the ways behavioral differences can show.

According to Dr. David Lykken, professor of psychiatry at the University of Minnesota in Minneapolis, "about 10 percent of identical twins have different dominant hands."[4] Dr. Lykken, who has conducted extensive research on the similarities and differences of twins, says: "Nobody knows the causes of handedness. Right-handedness tends to run in families as does left-handedness. But this is not the whole story, because many identical twins have different handedness." He adds: "Other factors are presumed to be involved, such as, perhaps, the position in the womb, but there are no definite answers."

Fraternal twins develop from two fertilized eggs and are no more genetically alike or different than any other siblings. Identical twins develop from a single fertilized egg that splits to form two embryos and eventually two babies. Dr. Lykken maintains that when this split occurs ten or more days after conception—after there has already been some cell division—"mirror twins" result, identical twins that are in some ways mirror images of each other. "This happens," he explains, "because the developing embryo has begun to develop laterality, with each side a little different."[5]

IF YOU COULD harness the power used by your brain, you could power a ten-watt lightbulb.

WHEN WE touch something, we send a message to our brain at 124 miles an hour.

Why can't I tell my left from my right?

Welcome to the club—there are many members! If you often have trouble telling left from right, you may have what is termed "handedness confusion." Approximately one-third of adults experience difficulties and frustrations in distinguishing their left from their right in common, simple daily activities.

Normally, children learn to discriminate between the left and right sides of their body at about age six or seven. Older children then expand this ability to include the left and right sides of other objects. For adults, this ability is essential to performing many ordinary activities, such as driving and reading.

If an adult does not master the skill—and this seems to be all too often the case—problems of confidence and competence emerge. Research shows that handedness confusion can last a lifetime. It is difficult to correct in adults, but less so in children.

Charles McMonnies, a Sydney optometrist and visiting professor of optometry at the University of New South Wales, claims that "many children with learning difficulty, especially dyslexia, have persistent problems with right-left discrimination, including reversals of letters and words."[6] McMonnies says, "It is important to teach children left-right body awareness early, not just to prevent problems in adulthood, but also to facilitate many aspects of early school experiences."[7]

Do animals ever show "handedness"?

According to Dr. Victor Denenberg, professor of biobehavioral sciences and psychology at the University of Connecticut, many animal species exhibit handedness (or is it "pawedness"?). He adds that, as in humans, handedness in animals is clearly under brain control. But unlike humans, a group of animals will normally split about fifty-fifty as to right-handedness or left-handedness. Furthermore, some research with nonhuman primate indicates that many use the left hand for simple tasks but the right hand for more complicated manipulations.[8]

Research by paleontologists Dr. Loren Babcock of Ohio State University and Dr. Richard Robison of the University of Kansas shows that ancient trilobites living 550 million years ago showed a handedness of sorts. Although these simple creatures had no hands, bite marks on fossils suggest an ancient propensity for turning to the right.[9]

Who lives longer, right-handers or left-handers?

One of the most interesting debates among handedness researchers concerns whether or not handedness is linked to life expectancy. In 1988, Dr. Diane Halpern of the California State University at San Bernadino and Dr. Stanley Coren of the University of British Columbia presented findings that left-handers had shorter life spans.[10] But in 1989, Dr. Max Anderson of the Canadian Statistical Analysis Service in Vancouver reported that left-handers had *longer* life spans.[11]

Then, in 1992, Dr. Charles Graham and colleagues at the Arkansas Children's Hospital in Little Rock concluded that left-handers do indeed live shorter lives—because, in an admittedly right-handed world, left-handed children have more accidents, some of which are fatal.[12] For example, equipment and mechanical devices are not designed with the left-handed minority in mind—and that often has disastrous consequences.[13] Another instance was presented by science commentator Daniel Bristow, who wrote: "An interesting example is the SA-80 assault rifle. When fired from the left shoulder, it ejects spent cartridges, at great velocity, into the user's eye."[14]

But according to Dr. Clare Porac, professor of psychology at Penn State University in Erie, "left-handedness is not necessarily the kiss of death." She notes: "Although the percentage of left-handed people among those over age sixty is lower than in the rest of the population, there is no indication that left-handedness leads to an early demise. Rather, a complex combination of factors combine so that fewer of the old and oldest old report left-handedness." One of those factors is that 80 percent of left-handers over age seventy-five were pressured to change handedness preference when they were children, with mixed results.[15]

Despite this, there is good news for "lefties" in another domain of behavior: memory. Research shows that "coming from a family full of lefties" tends "to make a person better at remembering events." That was the conclusion in two experiments conducted by Drs. Stephen Christman and Ruth Propper of the Department of Psychology at the University of Toledo.[16]

Are we right-headed or left-headed, just as we are right-handed or left-handed?

This question comes up from time to time, especially in connection with controversies about the right and left hemispheres of the brain and their separate functions.

Medical and behavioral science evidence suggests that we are "headed" just as we are "handed." That is, we are either right-headed or left-headed, much the same way that we are right-handed or left-handed.

Contrary to popular belief, the human skull is irregular in shape, oval, and thus asymmetrical. According to Dr. Grange S. Coffin of the Department of Pediatrics at the University of California Medical Center in San Francisco, there are two common patterns of asymmetry of the human skull, and one of these patterns appears much more frequently than the other.[17]

In the more common pattern, which Dr. Coffin calls "left-headed," there is a left-side and right-frontal prominence. It is as if "left-headed" persons had placed their hands on the sides of their head and twisted the left half of the skull backward and the right half forward. In the less common pattern, which Dr. Coffin calls "right-headed" or "reverse," there is a right-side and left-frontal prominence.

According to Dr. Coffin's figures, seventeen out of every twenty people are "left-headed." Moreover, he claims that "headedness" is determined very early in life—in fact, before birth. He believes that head shape is perhaps "embryonically reflecting" the shape of the mother's uterus, the mother's posture and sleeping position, the site of the embryo's implantation, or even "obscure tidal and gravitational forces."

THE AVERAGE human brain has about one hundred billion nerve cells.

ACCORDING to authorities on IQ measurement, most people have an IQ in the 90-to-110 range. You're considered a genius if your IQ is 132 or above.

Why is "water on the brain" so bad if the brain is normally surrounded by water?

"Water on the brain" is a very common birth defect and condition. Yet evidence suggests that there is much public ignorance about this condition, and many misconceptions surround it.

"Water on the brain," or hydrocephalus, is not a rare condition at all. In fact, it occurs in about one in every five hundred births. As such, it is more common than such birth defects as Down's Syndrome (about one in seven hundred births), spina bifida (about one in one thousand births), or cystic fibrosis (about one in two thousand births).

The so-called "water" of "water on the brain" is really not water at all but cerebrospinal fluid, the liquid that cushions and protects the brain and spinal cord from shock. This fluid is produced in the cavities of the brain called the ventricles. Normally, this cerebrospinal fluid flows continuously through the ventricles, bathes the surfaces of the brain and the spinal cord, and is absorbed into the bloodstream.

In hydrocephalus, however, this flow of fluid is seriously interrupted. The cerebrospinal fluid becomes trapped in the ventricles and therefore is unable to enter the bloodstream. The excess fluid causes the ventricles to expand, with the result that the brain becomes abnormally large. As a direct consequence of this larger brain, added pressure is placed on the developing skull, and the baby's "soft spot" (fontanelle) expands abnormally as well. If the pressure is not relieved

WOMEN'S brains are smaller than men's brains by an average of 12 percent.

early, permanent brain damage can result and the head may be horribly deformed.

More than 50 percent of all cases of hydrocephalus are congenital—developing before birth. These cases probably are associated with infections that affect the fetus. Nevertheless, hydrocephalus can also begin during birth as the result of birth trauma, or even during childhood as a complication of meningitis—an inflammation of the membranes that cover the brain and spinal cord (the meninges). In older children and even in adults, "water on the brain" can be triggered by a brain tumor or head injury. Often, hydrocephalus occurs along with other birth defects. For example, nearly 70 percent of children born with spina bifida also develop hydrocephalus.

One misconception about hydrocephalus is that it is always hereditary. Although there are occasions when the disease occurs twice in one family, the chance of this happening is no more than one in twenty. Another misconception is that there are proven ways to prevent hydrocephalus in the fetus. Although there are no proven ways as such, there is some indication that the risk of such birth defects as hydrocephalus and spina bifida may be reduced by taking certain vitamins. For example, the 1991 newsletter of the Guardians of Hydrocephalus Research Foundation in New York carried a reprint of an article claiming that vitamins, particularly folic acid, may minimize risk. Since 1992, the U.S. Public Health Service has recommended that women of childbearing age who wish to become pregnant take 400 mg of folic acid per day to lessen their risk of giving birth to a baby with birth defects.

Hydrocephalus can be detected through prenatal diagnostic techniques, such as ultrasound. When it develops in the days, weeks, and months after birth, it can be detected through head measurements, skull X-rays, and CAT scans.

If diagnosis and treatment occur early, normal mental and physical development is the most likely outcome, but if left undiagnosed and untreated, hydrocephalus can result in mental retardation, gross abnormalities, blindness, seizures, and muscle coordination problems.

Hydrocephalus is most commonly treated by surgically inserting an artificial tube called a "shunt." One end of the shunt is inserted into the ventricles, and the other end is inserted into another part of the body, usually the abdomen. This allows the cerebrospinal fluid to drain off and be absorbed into the bloodstream.

Shunt surgery, which in a child is normally conducted by a pediatric neurosurgeon, can relieve the worst symptoms of hydrocephalus and halt further brain damage or head growth, but it cannot reverse brain damage that has already occurred. Further surgery may be needed, for the shunt must be lengthened periodically as the child grows. The child will probably always need a shunt. Complications are few, but occasionally there are infections and shunt malfunctions, which, if serious, must be treated and corrected.

In the rare instances where "water on the brain" occurs in an adult, the same diagnostic and treatment techniques are employed.

Can I get brain damage by simply drinking too much water?

The kidneys can cope with only so much water. Theoretically, if you continually drink so much water so that your kidneys can't deal with the flood, your various body tissues, including brain tissue, can swell from the overflow, a condition called edema. You could eventually die if you continue to flood your body with water. You would be literally drinking yourself to death.

Can I easily put my thumb through a baby's "soft spot" and touch the brain?

Let's knock this one on the head right away, so to speak. A baby's "soft spot," located above the forehead, is called the fontanelle. Although a newborn's skull is not well formed at birth, the tissue covering the "soft spot" is very, very tough.

Can memories be injected from the brain of one person into the brain of another by using a hypodermic needle?

It is theoretically possible to transfer memories from one person to another—by injection. In a number of successful laboratory experiments beginning some fifty years ago, so-called "memory molecules" were successfully transferred via injection from one organism to another. Although the experiments so far have only involved simple life-forms, the implications for human learning are immense.

Imagine acquiring the works of Shakespeare, Einstein, or last night's homework assignment, not through painstaking study but from a hypodermic needle. And should memories fade with age, you could retrieve them from a vial as easily as a diabetic takes insulin.

ONE PART OF THE brain is called the Convolutions of Broca.

The memory transfer experiments began in 1953. In that year, Drs. Robert Thompson and James McConnell, then graduate students in psychology at the University of Texas, attempted to teach common planarian flatworms the simple task of avoiding an electric shock by swimming toward a light.

The planarian is a very simple creature and a favorite in laboratory experiments. It is a little over one inch long, found virtually everywhere in the world where there are ponds and streams, and the simplest animal to possess a true brain and a synaptic-type nervous system. Important too is that the planarian reproduces both sexually and asexually. Sexually, one planarian may mate with another planarian and subsequently lay eggs to produce offspring. Asexually, a planarian may merely split in half. In the latter case, both the head and the tail regrow to form two complete and genetically identical planarians. Each is at the same time its own mother, father, and twin sister and brother.

Just as with Pavlov's dogs, the Thompson-McConnell experiments showed that planarians could "learn" by simple stimulus-response conditioning. By 1955, Thompson and McConnell confirmed the then emerging theory of synaptic memory storage in the brain.

It had been theorized that memories are held somewhere within the synapses of the brain. A synapse is an extremely narrow fluid-filled space between two nerve cells. Transmitter chemicals released by the end of one nerve cross the synapse and stimulate another nerve to fire. In humans, this process is not only the foundation of brain activity but also the basis of every human thought. Swedish biologist Dr. Holger Hyden had further theorized that, within synapses, RNA (complex genetic molecules) actually held the memories. His experiments showed that the brains of trained rats were chemically different from those of untrained rats—hence the notion of "memory molecules."

In 1956, Dr. McConnell, now at the University of Michigan from where he retired in 1996, began memory transference experiments with another team of researchers. He, along with Drs. Daniel Kimble and Allan Jacobson, first trained planarians to avoid the shocks by swimming toward the light. Next, they cut them in half. After the heads and tails regrew into complete planarians, they then retested both halves and discovered that *both* the heads and the tails retained the memory of how to avoid the shocks. Thus, it seemed that memory molecules not only exist but also exist in multiples and can "migrate."

In 1960, Dr. McConnell, along with Drs. Reeva Kimble and Barbara Humphries, "trained" some planarians to avoid shocks, chopped them up into tiny bits, and fed them to another group of untrained planarians. A second group of untrained planarians received only normal food. Both "untrained" groups were then tested with uncanny results. The cannibal planarians retained the memory of the training received by the "educated" planarians they had eaten, but the noncannibal planarians showed no evidence of any learning.

Subsequent McConnell team experiments involved merely injecting RNA of trained planarians into untrained ones to achieve the desired "learning."

From 1964 to the present, animal experiments in the United States, Denmark, the Czech Republic, and elsewhere reproduced

MORE THAN 2,500 left-handed people a year are killed from using products made for right-handed people.

RIGHT-HANDED people live, on average, nine years longer than left-handed people.

these memory transference via injection results. Nevertheless, scientists still debate whether memories are encoded in RNA. Some dispute that memories were truly transferred from one animal to another in the experiments; others point out that replicating such experiments is not always successful.

Recent experiments, such as those by Dr. Joseph Farley of Indiana University, pinpoint the true "memory molecule" as protein kinase C (PKC). Protein kinase C seems to prime neurons to react strongly to a new stimulus. This line of work is being continued with a variety of researchers, such as Drs. Terry Crow and James Forrester of the University of Texas at Houston with snails, and Dr. Aryeh Routtenberg of Northwestern University with chicks.

Of course, the animals and the learning tasks involved in these experiments are primitive. There is a great leap from a worm learning to avoid an electric shock to a human learning a Shakespeare play. But what is fascinating about the memory transference experiments is that memories themselves emerge as real, physical things—capable of moving from being to being.[18]

Can the human brain be fitted with an implant to give it the ability to directly access computer data banks?

The day of this possibility is definitely on the horizon. As technology advances, the human body will incorporate nonhuman elements. Humans may soon become composite beings—part biological, part mechanical, part electronic. As one scientist puts it, humans will become "metamen."

Imagine surgically installing a tiny computer in your brain. A computer would be able to interface with your memory and thought processes and gain access to entire libraries of information. You may one day be able to carry around in your head the contents of the Library of Congress.

If you think this is science fiction or idle fantasy, think again. Technological innovations are bringing this fantasy closer to reality. Dr. Gregory Stock, a biologist and physicist in Princeton, New Jersey, argues that the human composite beings will be able not only to computer-to-brain access entire libraries but also to directly stimulate their brains in other ways. Mental problems will be a thing of the distant human past as humans will be able to calm themselves, concentrate their attention, or feel pleasure, all with the help of computer implants.[19] Furthermore, these future human brains will be so powerful that their owners will view today's greatest geniuses as no more than simpletons.

Dr. Stock maintains that the rest of the human body will also be transformed through its merging with machines: "There will likely be not one, but many 'human' forms in our future. As humans become more engineered, why would we not begin to manifest the same level of diversity seen in clothes, cars, and other designed objects? . . . By the standards of the future, a multiethnic society . . . will seem extremely homogeneous."

When will all this begin to happen? Dr. Stock says that the beginning is already here. Computer technology is moving so fast that brain implants will be commonplace by the year 2010. "Brain implants to boost thinking will be no more unusual than today's ear implants to boost hearing."

According to Dr. Stock, as metahumans become the rule rather than the exception, our entire civilization will be revolutionized for the better in a way that is unprecedented in human history. He observes that barriers will tumble down as petty differences between people and nations will be rendered insignificant by the international community of advanced "postbiological humans," and adds: "There can be little question that the middle of the next millennium will reveal a Metaman, not battered and tenously clinging to life as commonly portrayed in our twentieth-century postapocalyptic films, but rather healthy and growing, with human society thriving and the weighty problems of the year 2000 long since solved."

While Dr. Stock shows how machines can be inserted into the human, Dr. Hans Moravec shows how humans can be inserted into machines. Dr. Moravec, director of the Mobile Robot Laboratory at Carnegie Mellon University in Pittsburgh, has explained in detail how people could become robots, how they could download themselves into computers, and how all this could be accomplished within the next fifty years.

IT IS ESTIMATED that on an average day the human brain produces seventy thousand thoughts.

According to Dr. Moravec, this truly human robot á la computer "could now carry on the life of the person whose mind you transferred to it. The robot would have all the same skills and all the same motivations as the human being did, and so it could raise children or do anything else the human could do. In fact, and for all practical purposes this 'robot' *is* the human being. . . . *Everything* the human being did, this artificial replacement does too. So if you don't want to call it a human being, it seems like just perversity on your part."[20]

Whatever the form, Dr. Stock believes that present biological humans will eventually learn to accept, even welcome, the advent of postbiological "metaman." He notes that although social turmoil, uncertainty, and poverty will not disappear from the human landscape in the next few years, "humanity's trajectory is toward a rich and vital future."

He never "metaman" he didn't like.[21]

How do you measure the size of the brain?

The size of the brain is measured today by high-tech scanning equipment, such as CAT scans and MRIs. The measurements are precise. Before our modern age, it was impossible to measure a living person's brain precisely, but it was a different story if the person was dead. In that case a craniophore could be used.

Invented in 1895 by a U.S. Army surgeon, J. S. Billings, the craniophore held the deceased person's de-skinned skull in place so that measurements could be taken. Cranial volume and hence brain size was estimated by packing the skull with sand and then measuring the amount of sand needed to fill the skull.[22]

Will there someday be an artificial brain?

It is already possible to replace a part of the brain.

Dr. Theodore Berger and colleagues at the University of Southern California have developed the first brain prosthesis: an artificial hippocampus. The artificial hippocampus is in the form of a chip that is inserted into the brain and functions as a real hippocampus.

The hippocampus plays a role in memory. Researchers theorize that this artificial brain implant might force some people to remember things they'd rather forget. Experiments will progress from those involving laboratory rats to those involving monkeys and finally to those involving humans. Memory skills will be taught that require utilization of the implant. The researchers hope to have the brain insert available for humans who have suffered brain damage due to stroke, epilepsy, or Alzheimer's disease.[23]

chapter 3
the **head**

"**E**mpty heads and tongues a-talking, Make the rough road easy walking," wrote poet Alfred Edward Housman (1859–1936) in his "Shropshire Lad." No, Housman was not predicting the plots of television soap operas. But as he mentions heads—the subject of this chapter (we each have one, you know)—his lines provide a somewhat awkward introduction to what lies below.

Does a bigger skull mean a brighter person?

People have been asking this question for more than two hundred years, and many myths surround it. An entire field of pseudoscience called phrenology was built on the contention that skull size, shape, and bumps determined intelligence, personality, and even one's position on the evolutionary tree. Phrenology reached its greatest level of popularity in the late nineteenth and early twentieth centuries.

We can explode forever the myth that the larger a person's head, the more developmentally advanced that person will be. This was confirmed by researchers, headed by Dr. Teresa Brennan, previously of the Department of Pediatrics at the University of Virginia Medical Center in Charlottesville and now in private practice in Lynchburg, Virginia, examining the relationship between relatively small heads in children and later development.

The researchers used standardized developmental tests, including the Stanford-Binet IQ test (given at four years of age) and the Wechsler Intelligence Scale (given at seven years of age). The Brennan team found no developmental differences between the children with relatively small heads and the children with average or relatively large heads.[1]

Related research found that a child's narrow-headedness makes no difference in eventual developmental outcome either. The research team that made this finding, studying preterm infants, was led by Dr. Alison Elliman of the Queen Charlotte's Hospital for Women in London.[2] Too bad that head size does not correlate with intelligence. If it did, intelligence testing could be done with a tape measure.

Can you bore a hole in your head to relieve a headache?

This body mystery question arises for several reasons. First, when you have a headache the pressure builds inside your head.

People often fantasize that their head would feel better if the pressure was relieved. So why not a hole in the head? Second, it is a historical fact that the ancient Egyptians performed such operations to alleviate pressure on the brain due to stroke and other brain-based problems.

Let's look at this.

The most humorous report on "head boring" is found in the August 1986 issue of *People's Medical Journal.* According to this article, head boring, a fad long since thought to be dead and buried, has made somewhat of a reemergence.[3]

In 1962, Dr. Bart Hughes, an experimental neurologist in New York, proposed the unorthodox and highly suspect theory that a person's state of consciousness was dependent on the volume of blood in the brain. He advanced the idea that the brain was usually "constricted" by the skull and that, therefore, the best way to "raise consciousness" was to drill a hole in the skull. This is called self-trepanation.

Self-trepanation had a modest flock of true believers in the 1960s, and one disciple even wrote a book about various head-boring techniques. The author, W. Smith, titled the book simply *Bore Hole.*[4] (Perhaps it was meant to be a book on well digging?) Another follower, a young woman by the name of Amanda Fielding, gained a tiny slice of medical immortality of sorts by filming her own trepanation with an electric drill. Needless to say, head boring did not catch on in the 1960s.

But the head-boring fad may be making a modest comeback. Beginning in 1984, some U.S. doctors began receiving inquiries from patients about this form of "therapy." It also seems that some have been encouraged by the fact that head

REGIO FACIALIS is the technical name for the human face.

boring was mentioned in the enormously popular 1984 Hollywood movie *Ghostbusters.*[5]

Can you keep a severed head alive?

Medical science is now capable of keeping a severed head alive. In 1988, the U.S. government granted a patent for a perfusing device that would keep a severed head alive after being surgically removed from the body. According to the holder of the patent, with the device and today's drugs that eliminate blood clots and other waste products from the brain, a severed human head could be kept alive indefinitely.

The procedure calls for the head to be surgically severed from the body at the top of the neck, mounted upright, and attached to the perfusing device. The device consists mostly of plastic tubes connecting the bottom of the head and neck to circulation machines that continue brain maintenance.

Perfusion refers to the process of artificially keeping brain-sustaining oxygen, blood, fluids, and other elements circulating properly. Using the perfusion device would allow the brain to think, the eyes to see, the ears to hear, the eyelids to shut during sleep, and other head and brain functions to take place.

Chet Fleming, a St. Louis molecular biologist, engineer, and patent attorney, holds the rights to the perfusion device: U.S. patent 4666425. The patent, which was granted on the basis of blueprints only, is called a "prophetic patent" because there is no working model. Mr. Fleming plans to build his device and make it available to experimenters.

According to Mr. Fleming, writing in the *British Medical Journal,* "the technology for perfusing a severed head has important potential advantages, for research and for pro-longing life in a conscious and communicative state with, probably, less pain than many dying people suffer today. The difficult question is whether the advantages outweigh the disadvantages and dangers."[6] He maintains that the head-severing operation and use of his invention will definitely have some takers. "I have been contacted by half a dozen people who want to know how soon the operation will be available and how much it will cost," he says. "Some are dying; others are paralyzed. Most said that if the mind remains clear and the head can still think, remember, see, read, hear, and talk, and if the operation leads to numbness rather than pain below the neck, then they would want it."

In 1988, Mr. Fleming privately published a book titled *If We Can Keep a Severed Head Alive,* in which he outlines the reasons for his efforts on behalf of the perfusing device. One reason he obtained a patent is to protect the technology from falling into the wrong hands. Moreover, he promises to make his device available to "any scientist or surgeon who wants to try it" on either animals or humans—but if they do they must first consult three independent review panels, like those already existing in major universities.[7]

One panel would be an animal care committee, which would control experiments on animals. Another would be an institutional review board, which must approve experiments on human subjects.

IT IS TWO AND a half times easier to smile than to frown. It takes seventeen muscles to smile and forty-three muscles to frown. There are 656 muscles in the human body. Only a tiny fraction of these are needed to either smile or frown.

The final panel would be an institutional biosafety committee, which would control genetic engineering experiments.

Animal experiments in which the head is severed have been undertaken since the beginning of the twentieth century. In a notable 1907 experiment, the entire upper half of a dog was transplanted. For this work, the French physiologist Alexis Carrel received a Nobel Prize in 1912.

Perhaps the most famous experiment of this kind took place in the early 1970s. At a 1971 conference of surgeons meeting in New Haven, Connecticut, five doctors from Case Western Reserve University in Cleveland presented evidence that they were able to keep rhesus monkey heads alive for up to thirty-six hours. The leader of the team, Dr. Robert White, claimed that the monkeys remained fully conscious: "Our animals were not sleepy: they tracked; they ate; and they bit you if you brought your delicate hands near the mouth."[8]

According to Mr. Fleming, the White team's monkeys died "because of heparin overuse, a problem which can be overcome today using an extracorporeal heparin remover." Heparin is a blood anticoagulant. Mr. Fleming adds: "Research is being done today on intact brains which continue to generate brain waves after the sensory organs and the skull have been cut away."

The perfusion of severed human heads raises enormous ethical and legal, as well as medical, questions.[9] And if there are indeed those who would choose this operation, it proves once again how desperately humans will grasp at any hope of life continuing—in any form.[10]

Can you reshape your skull?

We all know we can change our mind, but less well known is that we can also change our skull—or at least others can do it for us.

Of course, the human skull in its normal, natural state is perfect for its purpose and needs no change. In fact, it is a marvelous piece of engineering. Contrary to popular belief, the skull is not one bone. Actually, twenty-two bones make up the framework of the human head—not counting the teeth. The average skull in adulthood is about eight and a half inches high, seven inches long, and six inches wide. The height of the skull is often used to estimate the measurements of the rest of the body. The eight sections of the skull that encase the brain, called the cranium, provide wonderful protection.

The skull also protects the eyes while maximizing our ability to see, and it protects the workings of the inner ear so that we could hear (although not as well) even if our outer ears were sliced off. The bones of the cranium have ragged edges that interlock with the edges of adjoining bones more finely than any jigsaw puzzle. These edges, called cranial sutures, follow no regular pattern and are as individual to each of us as our fingerprints.

There are many other interesting facts about the human skull, but the question remains: Why would anyone want to change this wonderful artwork?

A Sydney doctor was recently asked by a new mother, "Doctor, can you change the shape of my baby's head? It is *so* ugly!" The doctor reassured her that her baby was normal, that many babies are born with somewhat misshapen heads, and that, as a child grows, the head takes on a more pleasing shape in proportion to the rest of the body. In other words, there was nothing for her to worry about.

Nevertheless, his answer was unsatisfactory to the mother—she wanted her baby's head to be beautiful *now.*

In fact, there *was* something the doctor could do, although it would be unethical—and humans have being doing it for thousands of years. It's called head shaping, head molding, or "intentional cranial reformation." In the field of plastic surgery, it is called "nonoperative cranioplasty." In any case, it comes to the same thing: changing the shape of the skull to fit some desired image.

Head shaping by various means is a practice that "dates back to very ancient times and has enjoyed wide acceptance" in many societies, according to Dr. F. O. Adebonojo, a plastic surgeon from East Tennessee State University, writing in the *Journal of the American Medical Association.*[11] It is known to have occurred in Europe, Asia, and Africa and in both North and South America, but it was not practiced by ancient Aboriginal Australians. It dates back to at least 2000 B.C. on the Mediterranean islands of Cyprus and Crete as well as in ancient Egypt.

To successfully change the shape of the skull, the molding must be done in infancy, when the bones that make up the human skull are still soft and pliable. In most cultures where head shaping was commonly practiced, the molding process began within a few days after birth. External force had to be applied in some way. For example, among the Kwakiutl Indians of British Columbia, long heads were viewed as beautiful, so an infant's head was sandwiched between two slats of wood that were then tightly bound together with twine. This elongated the head somewhat. After a time, the head was rebound more tightly, producing further elongation. The process was continued for at least three months but sometimes much longer, until the desired degree of elongation was produced.

Among the ancient Incas of Peru, a naturally long head was made even longer, while a naturally short one was made shorter. Inca notions of beauty at the time dictated that natural characteristics, whatever they were, should be emphasized and exaggerated.

IT HAS BEEN determined that one brow wrinkle is the result of two hundred thousand frowns.

Sometimes people reshaped the heads of their children for reasons other than beauty. For example, in Tahiti and Hawaii only members of the royal class were allowed to shape their heads, which they did so that as members of the ruling class they could be easily distinguished from everyone else.

Sometimes, too, head shaping was employed for one gender only. For example, the famous ancient Greek physician Hippocrates (c. 460–377 B.C.) reported that female childhood head compression was used among the Greek aristocracy. It has been speculated that this was done both to make women more beautiful to men and to impair them intellectually—compression serving oppression. If so, then the head shaping of women among the ancient Greeks served much the same function as the foot binding of women among the prerevolution Chinese.

Dr. Adebonojo writes that, among societies practicing head shaping, "the range of intentional skull deformation (reformation) is quite large, varying from one extreme that produced grotesque cranial distortions to mild or minor deformations." Nevertheless, he adds, head shaping appears to be a fairly safe practice "since it does not appear to change cranial vault weight or volume or intellectual capacity."

Today, there is a definite place in modern medicine for various forms of head reshaping. According to Dr. Adebonojo, "some plastic surgeons have, in fact, considered the use of techniques similar to head shaping or molding as a valuable tool in the correction of certain craniofacial abnormalities." A classic article on this topic appeared in a 1973 issue of *The Lancet*.[12] Three New York University Medical School surgeons, led by Dr. F. Epstein, describe their treatment of five young patients with hydrocephalus: the treatment was "based on the premise that increasing the resistance to expansion of the skull by compressive bandaging promotes increased CSF [cerebrospinal fluid] absorption."

Furthermore, "nonoperative cranioplasty" is sometimes used to treat children with skull and facial birth defects of various kinds. For example, Hemifacial Microsomia/Goldenhar Syndrome (HMGS) is the second most common birth defect occurring in the Western world today. Approximately one in 3,500 births are affected. Usually the eye, ear, cheek, jaw, and one side of the face are deformed. Although HMGS is extremely variable from child to child, some children have cleft palate/lip, and organ and system problems, and a minority suffer mental retardation. Such children often require surgical re-formation of the skull. But the techniques today are somewhat more sophisticated than two slats of wood and twine.[13]

In any case, just as we humans have been known to change our minds, we also have been known to change our skulls.[14]

How do headhunters shrink a skull?

Headhunting is a curious custom found in many parts of the world. However, head shrinking is found only in some regions of South America.[15] Head shrinking is an elaborate process filled with cultural, symbolic, and religious significance.[16, 17] For example, when an enemy is killed, his head may be kept as a trophy for many years. Shrinking helps preserve the head. It may be believed that as long as the head is possessed, so is the enemy's strength, courage, and spiritual power.

According to Dr. Jim Leavesley, a retired general practitioner in Margaret River, Western Australia, "the result after hours or even weeks of dedication and ritual was a head about the size of a large orange," although techniques vary somewhat.[18]

After the enemy is killed, the skin of the neck is cut low onto the chest, the neck is separated with a sharp blade, and the head is removed. A strip of bark is then passed through the mouth, down the esophagus, and out through the end, to be used to hang or carry the head.

A pot is then filled with water and brought to a boil. The skin on the back of the head is slit up to the crown of the head and teased away from the bone so that it can be peeled completely forward. Eventually, the skull and lower jaw can be removed totally and are discarded.

What remains of the eyelids are then sewn together from the inside. Small pieces of wood are then put through the lips and the mouth is bound shut. A remaining mop of skin, scalp, and hair is dumped into the boiling water with special leaves for two hours. (Juices from the leaves supposedly keep the hair from falling out.) The skin over the back of the head is then brought together to resemble a deflated human head.

Next, rounded stones are heated in the fire, then levered into the severed neck opening. The head is then rotated by hand while another hot stone, held in a leaf, is used to smooth out the

THE HUMAN head contains twenty-two bones.

IT HAS BEEN
estimated that
banging your
head against
a wall every
ten seconds
consumes
150 calories
in an hour.

external features to give them a human appear-ance. The head begins to contract as it cools. Smaller stones and hot sand are dribbled into the head via the neck to permit all parts to shrink evenly. The hairs of the face, eyebrows, and eye-lashes are now disproportionately long, so they are singed off.

The head is then hung a little more than three feet above a fire, where it smokes overnight. Later, its natural oils are buffed to give it a per-manent luster, and the pieces of wood in the mouth are replaced with colored threads cut to the length of the hair. It is believed that if the mouth were to remain open the dead enemy could curse its slayer.

Why does the head of a baby chimpanzee closely resemble the head of a human infant, while an adult chimpanzee head is so different from a human adult head?

If you were to shave the face and hair off a baby chimp and keep the rest of its body wrapped up, it might pass for a human baby if you didn't look too carefully—but then again maybe not, because a chimp's ears are pretty big. In any case, baby chimps and human babies do look remarkably alike. It is also true that adult chimps and adult humans do not look at all alike. Why?

Human babies possess a round cranium with both a flat nose and jaw at birth. So do baby chimps. In fact, human and chimp embryos and fetuses look even more alike than they do as new-borns. The human newborn's brain grows rapidly, whereas the rate of brain growth for the newborn chimp starts to slow quite a bit. As a chimp grows, the jaw juts out, the nose stays flat, the teeth grow larger, and the superciliary ridge (the bone beneath the eyebrows) becomes more prominent. The cranial vault is lower and smaller than in humans.

According to the late Dr. Stephen J. Gould, famed paleobiologist at Harvard University, the difference between humans and chimps is that the human brain keeps growing at a faster rate for a longer time. The human skull must accommodate this larger brain—like the glove must fit the hand.[19]

chapter 4

the **eyes**

Poets call the eyes the windows of the soul. "Beauty is in the eye of the beholder" is attributed to Margaret Wolfe Hungerford (1855–97). "Night hath a thousand eyes" comes from John Lyly (1554–1606). "You've Got Reptilian Eyes (But I Still Love You)" is a song from Woody Allen's movie *Zelig* (1983). The immortal lines are endless.

So much has been written about eyes, yet most people know very little about them.

Can some people really pop their eyes out?

This is a weird ability, and very rare. The first case in the medical literature of someone popping his eyes in and out without any apparent injury or discomfort was reported in the *American Journal of Ophthalmology* way back in 1928.[1] Dr. H. Ferrer described a twenty-year-old male who "had the ability to dislocate at will either eye separately or both spontaneously." Four years later, Dr. J. Smith reported in the *Journal of the American Medical Association* that an eleven-year-old boy could do the same thing.[2] Every so often an article appears about someone able to perform this feat.

THE AVERAGE blink of the human eye lasts about one twentieth of a second.

The late actor-comedian Marty Feldman had eyes that *looked* as if they had popped out of his head, but he couldn't really pop them in or out. Actually, Feldman suffered from Crouzon's disease, which is sometimes called craniofacial dysostosis. People who have Crouzon's disease have eyes that look as if they are popping out of their sockets, but vision is not impaired. Feldman was no exception. On his passport where it asked for "any distinguishing physical characteristics" Feldman simply wrote "face."

Dr. Barnett Berman of Baltimore calls the ability to voluntarily propel the eyes "the double whammy syndrome." He writes about this weird ability: "I quote Emerson, who wrote, 'Some eyes have no more expression than blueberries, while others are as deep as a well, which you can fall into.' Once observed, the double whammy leaves an indelible impression that is unforgettable."[3]

Why do people cry?

Research suggests that we cry for physiological and emotional reasons. In fact, crying may be important to maintain physical and psychological health.

It is well known that crying is an emotional release and relieves pent-up stress, but what is not as well known is that the tears from crying are almost certainly one way for the body to cleanse itself of toxic substances. For example, salts are excreted in tears just as they are through sweat and urine. Tears contain a variety of different salts that come from the diet via the blood. Salt in food is absorbed by the intestines and enters the bloodstream. As blood flows through the tear-producing lachrymal glands, the salt enters tears.

French chemist Antoine Lavoisier (1743–94) conducted the first scientific study of tears in 1791, finding that tears contain sodium chloride (table salt). But tears also contain other salts, such as potassium chloride, plus other factors that assist salt formation. Among these are calcium, bicarbonate, and manganese. Experiments held more than thirty years ago showed that the concentration of sodium in tears was the same as that in blood.

There may be a strong element of truth in the expression "A good cry makes you feel better." The ancient Greek philosopher Aristotle (384–322 B.C.) theorized that crying at a dramatic performance benefits a person by a process called "catharsis"—the reduction of stress through emotional release. That word has figured its way prominently into our modern vocabulary of psychology. In a classic 1906 article in the *American Journal of Psychology,* Dr. Alvin Borgquist found that fifty-four out of fifty-seven patients reported positive health benefits after crying.[4] More-recent studies consistently report similar findings.

THE LENS of the eye continues to grow throughout a person's life.

THE OPPOSITE of cross-eyed is wall-eyed.

At the Ramsey Dry Eye and Tear Research Center in St. Paul, Minnesota, biochemist Dr. William Frey determined that "emotional tears" produced by tearjerker movies differ in chemical content from "irritant tears" produced by breathing onion-juice fumes. He found that "emotional tears" contain more protein than "irritant tears." However, the significance of this finding remains unclear.

In a study at the University of Pittsburgh School of Nursing, psychiatric researcher Margaret Crepeau found that among 137 men and women "healthy people are more likely to cry and have a positive attitude toward tears than are those with ulcers and colitis—two conditions thought to be related to stress."[5]

Researchers are currently investigating the content of tears for substances such as endorphins, ACTH, prolactin, and growth hormone, which are all released by stress.

Furthermore, the average cry lasts approximately six minutes. A typical one-year-old infant cries sixty-five times a month.[6]

Does any other animal cry?

The human being is the only primate that cries. Only one other land animal, the elephant, cries. Marine animals that cry include seals, sea otters, and saltwater crocodiles (the so-called "crocodile tears"). All these animals cry only to get rid of salt. However, one scientist, Dr. G. W. Steller, a zoologist at Harvard University who has studied sea otters extensively, thinks that sea otters are capable of crying emotional tears: "I have sometimes deprived females of their young on purpose, sparing the lives of their mothers, and they would weep over their affliction just like human beings."

Do women really cry more often than men?

It is often asserted that men rarely, if ever, cry—especially in public. The conventional stereotype is that crying indicates weakness in a man. The strength of this stereotype in forming opinion is exhibited from time to time. In 1972, U.S. Senator Edmund Muskie, then the leading candidate for the Democratic Party nomination for President of the United States, saw his hopes dashed when he was filmed sobbing while addressing a crowd. Senator Muskie later denied that he had cried. He attributed his tears to the near-freezing weather conditions. He may have been right, as exposure to intense cold can cause tearing, but the public refused to believe him. He was dismissed as "too weak to be President" and eventually dropped out of the presidential race and into obscurity—hardly anyone, pardon the phrase, shedding a tear.

In recent years, there has been some evidence that we are a little more likely to allow our male leaders to cry. The former Australian prime minister Bob Hawke is one example. First elected in 1983, he has cried in public on several occasions, yet he remains the second longest-serving prime minister in Australian history.

Nevertheless, evidence suggests that men cry more often than most of us are ready to admit. For example, surveys in the United States show that women average 5.3 cries a month, compared with 1.4 cries for men. That is nearly seventeen cries a year for the average man.[7]

If crying is a vital factor in health maintenance because it provides a release for emotions and stress, and if women cry more readily than men in our society, that may help to explain why men are more ravaged by stress-related diseases and die earlier. Perhaps women know that a tear a day keeps the doctor away?[8]

Why do tears taste salty?

As previously mentioned, tears contain salt. In fact, tears are approximately 0.9 percent salt. It would be impossible to mask the taste of salt. Anecdotal evidence from the behavioral sciences indicates that most people cannot distinguish the taste of tears from that of unpolluted seawater. Of course, finding unpolluted seawater nowadays is another matter.

Where do tears come from?

Above and slightly behind each eye, under the frontal bones of the skull, we have an almond-shaped lacrimal gland. About a dozen or so channels (lacrimal ducts) run from the lacrimal gland to the eye and eyelid. When we blink, the lacrimal gland is stimulated and tears wash over the eye. The eye is kept perfectly moist this way, and very clean. Tears are sterile and contain bacteria-destroying enzymes that protect against infection.

Where do tears go if not down my face?

When we cry, some moisture is lost through evaporation, but most of the moisture from tears drains from the inside corner of the eye, down through two other channels (lacrimal canals) into the peanut-shaped lacrimal sac, and finally into the nasolacrimal duct, where it drains into the nasal cavity. This is why your nose runs when you cry a great deal.

HUMAN BABIES born at term have open eyes and blink at birth. They probably do before birth as well, but it is difficult to detect.

Why do I sometimes cry at a happy ending?

Most psychologists maintain that it is a myth that we cry when we are happy. Actually, there is no such thing as tears of happiness. It is not out of happiness that we cry, but because unpleasant feelings are stirred up. Both men and women can suppress the urge to cry. During the emotional experience of a "four star, four hanky" movie, we may hold back the waters until the film's climax. We then have a climax of our own, so to speak: tears gushing forth. The energy used to hold back the flow is now discharged. These tears are an expression of this emotional release of combined feelings of anxiety, fear, sadness, relief, and so on.

It is not only at the movies that we seem to cry at happy endings. We emotionally release with tears when someone close to us is greeted after a long absence, when someone close emerges alive and well after delicate surgery, when someone close survives an accident, and at other times. Such emotional occasions are endless. And we never run out of tears.

Why do we blink?

Studies show that humans blink every two to ten seconds on average. We do this automatically, as our eyelids protect our eyes from injury. Occasionally, we blink voluntarily as part of our body language. For example, we wink or "bat an eyelash" in order to send flirtatious messages to another person. We also blink to express astonishment as if to say, "I can't believe my eyes!" Nonhuman primates, such as baboons, have a white region of the upper eyelid.

When they blink it is a threat gesture that signals both an alarm and a warning.

In humans, the inner surface of the upper and lower eyelids, as well as the eyes themselves, are covered by a thin and almost transparent membrane called the conjunctiva. The conjunctiva keeps the surface of the eye moist, for moist eyes are necessary for proper eye movement. During blinking, the eyelid spreads moist tears over the conjunctiva and also into the corners of the eye.

Why can't some people control their blinking?

Sometimes a person cannot control eye-blinking. In that case, they may have a medical condition called blepharospasms—uncontrolled eyelid movements.

TRUE BLACK eyes are not known in humans. Human eye color is a complex trait. Depending on the amount of different pigments of coloration in the human eye, color varies from very light blue to very dark brown.

No one knows what causes blepharospasms, how to cure them, or how many people are afflicted by them. Few doctors even knew about this condition a decade ago, but research is coming up with some interesting clues to understanding blepharospasms.

Dr. John Hotson, chief of neurology at the Santa Clara Valley Medical Center in San Jose, has long been interested in the way blepharospasm patients see and remember actions. He believes that the way the eyes see movement can serve as a clinician's window to pinpoint the defect at the root of blepharospasms. If he is successful, Dr. Hotson's work would be a step toward developing a diagnostic

EYES DO NOT grow very much from infancy to adulthood. The diameter of the orb (eyeball) is about 0.68 inch at birth, 0.79 to 0.83 inch at puberty, and 0.94 inch in adulthood.

technique and cure for this bizarre ailment that turns patients' lives upside down.

Because eyes normally blink once every few seconds, the lids close so gently and quickly that vision is not interrupted at all. In many people, an irritation or a tic may cause the lids to flutter or shut unusually hard. However, "blefros" (as blepharospasm patients are sometimes called) suffer spasms of a different kind. Their eyebrows yank down, and their lids jerk and slam shut. Over time, the spasms spread. The jaw may lock, the face may twitch, the neck may be gripped in pain.

Although such patients are often rudely dismissed as having a psychiatric-based problem, blepharospasm experts, such as Dr. Hotson, are convinced that the problem is physical and not psychological. This seems to be confirmed by his research, which suggests that patients reflect a typical range of personality types and perform normally on psychological tests. Moreover, according to Dr. Hotson, the ailment progresses in predictable patterns similar to the patterns doctors often see in patients with Parkinson's disease.

Currently, two operations can ease the symptoms of blepharospasm. In one, the surgeon makes an incision near the ear, pulls down the forehead, and severs the branches of the nerve that controls eyelid movement. The procedure is delicate and often disfiguring. The second, a more common surgical option, involves surgical extraction of the squeezing muscle in and around the upper eyelids.

Many blepharospasm patients rely on experimental injections of botulinum toxins, the same poisons that cause botulism. Nearly three hundred doctors worldwide are testing this therapy. Patients receive a series of injections in their lids and brows to paralyze the muscles. While the injections appear to be effective, they wear off after only a few months and must therefore be repeated.

Dr. Hotson hopes to find a longer-lasting remedy. As a first step, he is searching for the cause of blepharospasms. He theorizes that wild blinking reflects miscommunication among nerve cells located in the brain's basal ganglia (a major center for controlling movement). The effect the basal ganglia pathways have on eye movement is currently receiving Dr. Hotson's closest attention.

According to Dr. Hotson, research shows that a pathway abnormality can subtly hamper a person's memory for movement. For example, although a person could watch a dot shoot across a computer screen and believe that he or she recalls the directions of movement perfectly, that person would probably be unable to trace the line of movement as accurately as someone without this brain disorder could.

Dr. Hotson is looking for evidence of such an impairment in blepharospasm patients by having research subjects watch dots dance on dark screens and carefully monitoring their speed and accuracy in "charting." The research will continue for a few more years at least before firm results can be seen. After that, we shall see.[9]

Am I happier the more often I blink?

Medical or behavioral scientists do not make much of this, but it is possible that the happier we are, the more often we are likely to blink.

This is the firm assertion of one noted U.S. neuroscientist, Dr. Joseph Teece, a professor of neuropsychology at Boston College. Dr. Teece claims that research shows that eye blinks offer clear signs of emotional well-being: "Pleasant feelings mean fewer blinks, while negative ones such as anxiety and pain get those eyes afluttering."[10]

To illustrate his point, Dr. Teece claims that during the 1988 U.S. presidential debates between George Bush Sr. and Michael Dukakis, "George Bush's average rate of 67 blinks a minute went to 89 when talking about abortion and to 44 when praising Dan Quayle." According to some body language experts, eye-blink rate is a predictor of who will win an election where television image is critical. For example, in the 1992 U.S. presidential debates, Bill Clinton had a somewhat lower eye-blink rate than George Bush, but the difference was small—as was Clinton's margin of victory.

Eye blinks are already being used in health settings. For example, some dentists are using eye-blink rates to track when patients are feeling pain, and psychotherapists are using them to tell when patients are uncovering painful emotions.

Blinking may have medical diagnostic uses. According to Dr. Craig Karson, a psychiatrist in Little Rock, Arkansas, eye blinks will soon help diagnose brain disorders, anxiety, and depression, just as they now help to diagnose schizophrenia and Parkinson's disease. For many years it has been observed that patients commonly show irregularities of eye-blinking in both schizophrenia and Parkinson's, but only recently have these patterns of irregularities been understood.

In any event, research on eye-blinking is worth keeping an eye on—if for no other reason than to read the moods of others. As Dr.

Teece adds, "If your current flame has eye blink storms when professing love, watch out."

What causes baggy eyes?

There are actually rings around the eyes as well as bags under them, but the bags look worse, and that's why we worry about them more. According to Dr. Bette Albert, previously of the School of Public Health at the University of California at Berkeley and now at the Johns Hopkins University School of Medicine in Baltimore, bags and rings have the same cause: they are the result of a body-fluid problem that appears as puffiness. "Bags" may occur when fluid accumulates in the area under the eyes, where the skin is thinner than anywhere else on the body. With advancing age, and perhaps with the assistance of hereditary factors, puffiness may become more prominent or even permanent because skin gradually loses its elasticity and may begin to sag.

Other factors may be involved as well. Some individuals have permanent bags that may be due to a specific hereditary condition in which the fat that cushions the eyeball protrudes through weakened muscles. Certain medications, such as cortisone, and allergic reactions to cosmetics, tobacco smoke, or air pollution, may aggravate the situation.

Generally, eye fatigue or irritation may make eyes puffy. Thyroid, kidney, or heart disease can

WHEN YOU focus your eyes, small muscles in the eye (ciliary muscles) contract and adjust the shape of the lens. The process of focusing modifies the ability of the lens to refract light.

THE HUMAN EYE is capable of differentiating ten million colors.

also increase fluid retention, which can be particularly noticeable around the eyes. And if this were not enough, another factor in the baggy-eye picture is simply the force of gravity. While sleeping, especially on the stomach, extra fluid may pool in the upper and lower eyelids.

Besides avoiding the factors already mentioned that worsen the problem, there is not much one can do about the puffiness. Sleeping with the head elevated on an extra pillow may be enough to allow gravity to drain the eye area. In severe cases, the sagging tissue or excess fat under the eyes can be removed surgically, frequently on an outpatient basis.

What causes dark circles under the eyes?

Dr. Bette Albert accounts for the phenomenon of dark circles under the ideas with much the same explanation. Again, age is a major factor. Dark circles under the eyes also tend to be a family trait and to worsen with age. However, they are seldom a symptom of any underlying medical problem.

What appears as a dark or blue-black tint is actually the blood passing through veins located just below the surface of the skin. These circles may be darker when the eyes are tired. Dark circles under the eyes often occur in women during menstruation or during pregnancy.

Cosmetic concealers will cover up the problem if regular makeup is insufficient.[11]

Why do I squint?

The glib answer is that we squint in order to see better. The real question is, why does squinting improve vision?

Without going into a far too technical description, the eye receives rays of light and bends them so that an image is resolved on a small point of the retina. But things can go wrong. If the rays focus in front of the retina, the person has nearsightedness (myopia) and suffers blurred vision of distant objects. But if the rays focus at a point behind the retina, the person has farsightedness (hypermetropia) and suffers blurred vision of nearby objects.

According to Dr. Stephen Miller, director of the clinical care center of the American Optometric Association in St. Louis, "the shape of the eyeball and the focusing power of the lens and cornea help determine focus, but the angle at which light rays hit the eye plays a role." He adds, "Light comes into the eye from all directions, [and] rays entering the eye at an angle from above or below would tend to focus somewhere before or behind the center of vision. . . . Those rays coming in essentially perpendicular to the eye, on the other hand, would tend to be focused more directly on the retina, providing a clearer image of what one is looking at."

Therefore, according to Dr. Miller, "the basic impact of squinting is to reduce the number of superficial or peripheral rays of light that enter the eye, so only the rays coming directly in are focused on the retina. . . . This cuts out a lot of the rays that are out of focus and eliminates a lot of what would otherwise be a blurred image."

Dr. Miller observes: "[You aren't] going to solve vision problems by squinting your way through life, but squinting might help someone who has lost his glasses and needs to see a road sign."

Too much squinting can result in headaches and "squint lines" in the face. It is probably a good idea to see an eye-care professional if you find yourself squinting much more often than before. Sometimes others will point this out to you. Vision correction is far more preferable than continued squinting, in the long run—taking a farsighted view, of course.

What causes a sty?

D r. Bette Albert provides simple answer again: A sty is basically a pimple on your eyelid caused by an infection in the follicle from which an eyelash grows, or in a connected sebaceous (oil-producing) gland. In order to combat this infection, the body musters an additional supply of blood to the area, which brings the body's infection fighters into the front lines of the battle, but it also produces a sty, "a red painful inflammation resembling a small boil."

Dr. Albert adds: "In most cases a sty doesn't require special treatment. Within a few days it will come to a head and burst; the inflammation should subside in a week or so. You can hasten the process—and relieve some of the pain—by applying warm compresses for fifteen minutes three or four times each day." However, Dr. Albert is quick to caution, a sty should never be squeezed, nor should the eyes be rubbed, because either can spread the infection. Instead, it is recommended that "once the pus has come to a head, you can speed up healing by carefully pulling out the lash from the infected follicle."

Moreover, Dr. Albert observes, "an occasional sty is one of the more common eyelid problems," but "if you get sties frequently . . . or if they are severe and resist healing, see your doctor. If [a sty] is very large, your doctor may prescribe a med-icated ointment."[12]

Why do I sometimes see spots before my eyes?

One *could* ask what you'd been drinking or smoking. But in all seriousness, strange spots before our eyes are called "floaters." These spots can also appear as dots, circles, lines, or even complicated geometrical patterns. Floaters are not outside the eye or even on it, but floating *inside* the eye. They are caused by small pieces of tissue and debris drifting through the eyeball's gelatinous filling, called the vitreous.

A normal aspect of the aging process, these particles can block light rays traveling toward the retina, the eye's membrane, which receives visual images and communicates them to the brain via electrical impulses. The resulting shadows are seen as spots drifting through the field of vision. Floaters are especially conspicuous against plain backgrounds, such as a white wall or a blue sky.

The time of day can often affect the appearance of floaters. They seem to be particularly apparent in the morning upon awakening, especially if one stares up at a blank ceiling. When a person is lying horizontally, the tissue and debris settle down toward the part of the retina that registers "straight-ahead" vision. Blood cells and other infection and inflammation by-products in the eye can also cause floaters, but all are eventually absorbed by the body.

Floaters are normally harmless. However, according to Dr. John Kiefer, a San Francisco eye specialist, occasionally "floaters are a

ONE-QUARTER of the human brain is used to control the eyes.

THE IRIS IS THE membrane that controls the amount of light entering the eye.

warning of a more serious problem." He writes: "If you suddenly notice a marked change in the size or number of floaters, or the appearance of a pinkish-red floater, or floaters accompanied by a persistent flashing light, consult an ophthalmologist."

Dr. Kiefer observes that "most people get used to the everyday variety of floaters. But if they bother you, try moving your eyes quickly—side to side and up and down. By swirling the fluid in your eye, you may move the floaters out of your line of sight for a while."[13]

When a nearsighted person holds a mirror up close, why are distant objects still blurry?

Surprise, surprise about this one! The viewer is not actually seeing the distant object as if it were brought up close by the mirror. Dr. Clint Hatchett, an astronomer with the American Museum of Natural History's Hayden Planetarium in New York, says: "What you're doing with a flat mirror is that the person is in a sense seeing what is called a virtual image of an object that is visually the same distance away on the other side of the mirror"—that is, "it is as if he were looking right through the mirror" at the object. He goes on: "Similarly, with a camera, if you are photographing yourself in a mirror, you have to focus for the distance that is twice as far away as the mirror is." Dr. Hatchett has tested this himself by focusing for the distance of the mirror. "And it came out fuzzy," he says.

WHY DOES Mona Lisa have no eyebrows? That's because in Florence, Italy, during the Renaissance it was the fashion for women to shave them off.

What causes eye color?

The pigment of the iris gives the eye its color. This pigment is called melanin. The amount of melanin determines the eye color. Large amounts of melanin result in darker eyes (black, brown, or hazel). Smaller amounts produce lighter eyes (green or blue). People suffering from albinism have pink-colored eyes due to the absence of melanin, making the iris transparent. As a result, the blood vessels of the eyes show through.

The amount of melanin is determined by genetics. There are more dark-eyed than light-eyed people in the world because there is a genetic trait dominance toward more rather than less melanin. That is why when one parent has dark eyes and the other has light eyes, their offspring are more likely to be dark-eyed.

Why does eye color seem to change with age?

Eye color changing with age is largely an illusion.

It is not so much a question of the eye changing color, but more a question of the eye changing size—at least part of the eye. In fact, the color of the iris does not change with age. However, the pupil (the black hole at the center of the iris) may shrink with age, making the center of the eye less dark. Therefore, when the entire eye is observed it appears to be lighter in color overall.

Commenting on this illusion, Dr. Richard Thoft, a professor of ophthalmology at the

THE EYES OF A baby do not produce tears until the baby is six to eight weeks old.

University of Pittsburgh, says that, for example, bright-blue eyes may slowly seem to turn lighter as a person ages.

Why does the eye have a "blind spot"?

Each eye has a "blind spot" in the retina that corresponds to the head of the optic nerve. However, despite these visual black holes, no vision is lost. An old theory attempting to explain this phenomenon is that the brain simply ignores the visual blind spots. But laboratory experiments showed that this explanation was inadequate when scientists created artificial blind spots in the visual fields of subjects.

Research by Dr. Vilayanur Ramachandran of the University of California at San Diego and Dr. Richard Gregory of the University of Bristol suggests another view. The doctors' theory is that the brain compensates for these natural holes in the visual field by creating a physiological representation of visual information surrounding blind spots. This creation automatically paints a coherent scene by filling in the optical void.[14]

IT TAKES THE human eyes about one hour to fully adapt to seeing in the near dark. Once they have adapted, however, the eyes are about one hundred thousand times more sensitive to light than they are during daylight hours.

Why do I often look up while thinking?

It is not widely known, but most psychologists believe that people use either vision, hearing, or touching when they seek a mental solution to something. An image, sound, or feeling from the past, or a new one we construct, aids our thinking.

For example, when someone tries to recall the number of degrees in the angle formed by the hands of a clock at ten o'clock, they might first try to picture the visual image of the face of a clock, then focus in on the 10 and the 12, and then focus in on the big hand pointing to the 12 and the little hand to the 10 (30 degrees). In any case, a visual image that is based on the sense of sight is utilized.

Specially trained psychologists, known as neurolinguists, have theorized that specific eye movements indicate what sense is being relied on in any given thought. In fact, there are seven of these:

"Up-right" indicates visually remembered images.

"Up-left" indicates visually constructed new images.

"Straight-right" indicates auditorily remembered sounds or words.

"Straight-left" indicates auditorily constructed new sounds or words.

"Down-right" indicates auditorily sounds or words.

"Down-left" indicates kinesthetic feelings, which can include smell and taste.

"Straight ahead" indicates that information is being accessed.

If this theory is correct, we sometimes look upward when we are thinking, to draw on either an old or a new visual image to help us out.[15]

Can eating carrots improve my eyesight?

There may be some truth to this enduring notion, which is both one of the oldest old wives' tales and also one of the standards in the "Mom's useless advice" category. But Mom might be right about eating lots of carrots to improve vision.

According to Dr. Dennis Baylor, a professor of neurobiology at Stanford University, carrots are high in vitamin A, which is known to be essential for eyesight. A serious deficiency in vitamin A can cause night blindness—a condition known to improve when sufferers eat carrots. Nevertheless, if you are already getting enough vitamin A in your diet, eating carrots probably will have no effect on your eyesight.[16]

Vitamin A is an important ingredient in the body's manufacture of retinal, the light-sensitive chemical that enables cells in the eye's retina to react to light. When light strikes a molecule of retinal, it changes the molecule's shape. This activates a cascade of chemical events that send a message to the brain informing it that light has struck the retina. Dr. Baylor adds that the eye has two kinds of light-sensitive cells: rods and cones. Rods, the cells we rely on to see in dim light, are most affected by a vitamin A deficiency. In order to function properly, the rods require that every retinal molecule be in place and ready to respond to light.

Vitamin A deficiency can cause a retinal shortage in the rods, and with some of the retinal sites empty, the rods will never fully adapt to the dark. Hence, you will not be able to see well at night. Dr. Baylor believes that eating more carrots, or any other source of vitamin A, should alleviate the retinal shortage and restore night vision.

It's not just carrots that can help vision—spinach may do it too. Although there's no guarantee of attaining Popeye's strength by eating spinach, a diet rich in spinach can help ward off the leading eye disease of the elderly: geriatric macular degeneration. The macula is the part of the retina located at the back of the eye. In geriatric macular degeneration, the macula becomes progressively damaged. As months and years go by, vision gets worse and eventually all vision is lost. There is no cure.

IF WE ADD UP all the time we spend blinking, we spend about half an hour a day, or approximately five years of life, with our eyes shut while we are awake.

According to a 1994 study by Dr. Joshua Seddon and colleagues from the Massachusetts Eye and Ear Infirmary in Boston, a diet high in spinach results in a 43 percent lowered risk of developing geriatric macular degeneration.

Popeye was right, but for a different reason. And if Mom is getting older, tell her about the spinach findings to help her vision. While you're at it, thank her for her carrot advice. Tell her you're wearing clean underwear too, and you'll really make her happy.[17]

Can the eye suffer a stroke?

We all know that the brain can suffer a stroke, but what about the eye? Although the eye doesn't suffer the stroke, the optic nerve certainly can—and frequently does. When the optic nerve suffers a stroke, there is an abrupt loss of vision.[18]

Will television cause eye damage?

It is a common misconception that television causes vision damage. Although the contrast between a bright television screen and a dark room will temporarily tire the eyes, there is no long-term eye damage. Furthermore, there is no risk of eye damage from the reflective glare off the screen from a poorly placed lamp or other light source. Nor is there any need to fear that sitting nose-to-screen will cause nearsightedness or damage children's eyes in some other way. According to Dr. Theodore Lawwill, spokesperson for the American Academy of Ophthalmology, "some people with mild cataracts may even see the screen better in dim light."

Dr. Lawwill adds that children "like to be as close to the action as possible and would climb into the TV if they could." Nevertheless, young children are able to focus sharply on objects as close as a few inches away from their eyes. This distance lengthens as they get older. Thus, Dr. Lawwill claims, they "may block the screen for other viewers, but it won't hurt anybody's vision."

Dr. William Beckner, senior staff scientist at the U.S. National Council on Radiation Protection and Measurement in Washington, D.C., also dispels the notion that television causes radiation damage to eyes. Dr. Beckner says that, compared with television sets built twenty-five years ago, when his organization first warned of possible radiation risks, "modern receivers are built differently, using lower voltages and better shielding. No matter how close you sit to the set, X-rays just aren't a problem."[19]

PEOPLE WITH BLUE eyes are better able to see in the dark.

Is there really an "evil eye"?

Belief in the power of the "evil eye" to injure or kill still survives today. A person who believes in the evil eye is convinced that someone can harm someone else physically by merely looking at them in a strange way.

Belief in the evil eye in various forms has been found in most but not all parts of the world. In many places, the belief survives unchanged—as it has for centuries. Presumably, many immigrants bring it with them to their new countries. Some modern English expressions, such as "dirty look," "looking daggers," "stare someone down," and "if looks could kill," are remnants of belief in the evil eye.

The evil eye has been rarely studied in behavioral science, but the little research that does exist shows that we might be wrong in dismissing belief in the evil eye as belonging only to the superstitious traditions of others. In *The Mirror of Medusa*,[20] Tobin Siebers writes about a study that attempted to assess "the impact of staring in interpersonal psychology": "The [Stanford University] students surveyed, although educated and rational individuals, believed that the intensity of the eye could be detected. Of 1,300 students, 84 percent of women and 73 percent of men believed that a stare could be *felt.* Like many modern individuals, these students would have never taken the evil-eye superstition seriously, yet they unwittingly voice one of its fundamental premises."

We should not make too much of the Stanford survey. Nevertheless, we just might believe in the evil eye, or at least in one aspect of it, a little more than we think.

Historically, belief in the evil eye was widespread throughout Europe, the Middle East, and India, and among lesser-known cultures studied by anthropologists. It is interesting, however, that according to Clarence Maloney in his book *The Evil Eye*,[21] it was unknown among the oldest people on Earth, the Australian Aboriginals.

Virgil speaks of the evil eye making cattle lean. From its Latin name *fascinum*, we get the word "fascination." In ancient Rome, professional "evil eye sorcerers" were hired to bewitch enemies. By the Middle Ages, Europeans were so fearful of falling under the influence of the evil eye that people accused of looking oddly at others were often burned at the stake as witches. In the sixteenth and seventeenth centuries, hundreds of women were killed for this reason. Evidence against them consisted solely of accusations that someone had died or fallen ill after "being looked upon." During such witchcraft trials, even judges were often so afraid of the evil eye that the accused was led into the courtroom backward and made to face the wall. Crossed eyes (strabismus), any eye injury causing misalignment, or even a serious case of cataracts could lead to accusations of witchcraft resulting in trial and death.

Some historians have asserted that the practice of blindfolding prisoners at their execution was, at least in part, intended to protect witnesses from attempts of the condemned person to seek revenge by using the evil eye.

Several theories account for the origins and predominance of the belief in the evil eye. One suggests that it might come from the primal fear of early humans that being stared at by a predator foreshadows an imminent attack. The odd stares and angry glances of hostile neighbors, evil spirits, or jealous gods may often have been considered warnings of ill happenings to come.

Another theory holds that early humans must have found it frightening to glimpse their own image reflected, in miniature, in the eyes of someone else, and might have taken that as a sign of being in immediate personal danger. Perhaps the fear was that one's likeness might become lodged permanently within the eyes of another person, which would mean that one's soul had been stolen by the evil eye. Anthropologists, missionaries, tourists, and others have often remarked that, even in the twentieth century, natives in some parts of Africa believed that being photographed would result in permanent loss of one's soul—captured on film, as it were.

THE PUPIL OF the eye expands as much as 45 percent when a person looks at something pleasing.

IT IS IMPOSSIBLE to keep your eyes open while sneezing.

Supposed remedies for the evil eye are just as curious as the belief in the evil eye itself. The Congolese gave an evil-eyed male sorcerer beer and tobacco to placate him. The Scots tied red ribbons to the tails of livestock.

The ancient Egyptians had a curious antidote for the evil eye: kohl, history's first mascara. Worn by both men and women, kohl was applied in a circle or oval around the eyes. The basic chemical ingredient for this was the metallic substance antimony. If they could afford it, aristocratic Egyptians had a soothsayer prepare the compound and apply it, but women often added their own special, usually secret, ingredients to the antimony formula. Such concoctions were said to convey additional powers to the wearer.

Besides warding off the evil eye, ancient Egyptian mascara may have had an additional justification for use. Darkly painted circles around the eyes absorb sunlight and consequently minimize reflected glare into the eyes. The ancient Egyptians, living in the hot, harsh, sun-drenched climate of North Africa, may have discovered this fact.

But the ancient Egyptians failed to discover another antiglare device: sunglasses. The earliest sunglasses were devised by the fifteenth-century Chinese, and in southern China at the time, one use of sunglasses was to hide the wearer's expression in court—and to guard against the evil eye.[22]

What is color blindness?

Total color blindness (TCB) is rare, according to Dr. James C. Trautmann, an ophthalmologist at the Mayo Clinic in Minneapolis. Total color blindness occurs only when a person sees everything in shades of gray. More often, a person with poor color vision has difficulty seeing reds and greens and instead perceives colors as mostly blues and yellows. Yet this individual may learn to perceive red and green by recognizing varying amounts of brightness.

Dr. Trautmann adds that, as an inherited condition, color blindness is much more common in males than in females. While less than one percent of females are color-blind, the figure soars to 8 percent for males.

Furthermore, color vision can also be negatively affected by eye diseases, such as cataracts, optic neuritis, or retinal disease. But Dr. Trautmann claims that these cases account for very few instances of color blindness.

According to Dr. Trautmann, although inherited color blindness cannot be treated or cured, it is not a serious medical problem either. People who know they are color-blind can easily compensate for their condition. For example, when purchasing clothes, they may ask for assistance in color coordinating.[23]

the **nose, ears,** and **mouth**

Julius Caesar claimed Cleopatra's large nose was the key to her beauty. Ol' Juli baby supposedly had a very big one too. So did Leonardo, Galileo, Voltaire, and Jimmy Durante. But Edmond Rostand's Cyrano takes the prize. It's a sign of great dignity, provided you don't look down it at others too often, keep it out of other people's business, and don't get it out of joint.

Of course, we're talking about the nose here—and hopefully not talking through it. Instead, the nose is the stuff from which legends are made. And it doesn't have to be a stuffy nose either. Our nose is always a little bit ahead of us. It's as plain as the nose on your face.

But while we're at it, let's not forget the ears and mouth. What was Gable's fame was Dumbo's triumph. And while comedian Joe E. Brown reputedly had the largest mouth in Hollywood history, for many others we'll never know—their lips are sealed.

Can my nose grow bigger?

Remember the story of Pinocchio, the little wooden puppet who wanted so much to be a boy? Every time he told a lie his nose grew bigger. In fact, the Pinocchio effect is very real, at least to some extent.

The inside of your nose is classified as erectile tissue. It's just like your you-know-what or your you-know-what-else. Your nose gets slightly larger or smaller depending on blood flow. What you eat, temperature, illnesses, allergies—even emotional states, such as level of anger—can alter the size of your nose.

It is common for people to feel they have a temporarily stuffy nose after eating. Some even say they can feel it expanding or contracting. The size difference is tiny, but if you measure your nose throughout the day, you'd find that it does get smaller or bigger—but never as dramatically as poor Pinocchio's prominent proboscis.

How do people who are totally deaf from birth articulate their thoughts internally?

There have been arguments of all kinds about the need for language in thought, and vice versa.

The debate probably began with Aristotle, who argued that deaf people would never learn language or be able to think. Believe it or not, some people still hold this view. According to Dr. Howard Busby, a linguist and speech pathologist at Gallaudet College in Washington, D.C., "Some people still think that deaf people don't think. But deaf people do think and articulate their thoughts with a variety of languages ranging from English to American Sign Language."

Dr. Busby adds: "Most people visualize, in their mind's eye, so to speak. What actually makes them think that they're thinking in words is when they try to articulate it. The description of those thoughts gives language to them. The language you choose to describe them forms the shapes of those thoughts. If you articulate them in English, you assume that you thought them in English." To illustrate, Dr. Busby elaborates: "Think of a high jumper. He's standing back, he's looking at where he's going to jump; he doesn't tell himself, 'I'm going to spring on my right foot.' He doesn't really say it. He pictures it instead."[1]

Dr. Busby has a reasonable view, but many in psycholinguistics will no doubt dispute this.

Has anyone asked Aristotle lately?

THE EASIEST sounds for the human ear to hear, and also those that carry best when pronounced, are, in order, "ah," "aw," "eh," and "oo."

OF THE FIVE senses, the one that is less sharp after you eat too much is hearing.

ACCORDING TO experts, smell is the sense that is most closely linked to memory.

Why do I yawn?

Yawning is hardly a ho-hum topic. Research reveals there is little scientific evidence to back up many of our popular opinions about why we yawn, when we yawn, what function yawning serves, and what circumstances affect changes in yawning behavior.

Evidence suggests that yawning is triggered by as yet unknown physiological states. However, it is true that witnessing yawning can immediately provoke yawning—one of the few human behaviors where this occurs. In fact, yawning can be triggered by merely reading about or just thinking about yawning.

Such findings are the product of research conducted by the world's foremost authority on human yawning, Dr. Robert Provine, a professor of psychology at the University of Maryland. Dr. Provine and a few other intrepid researchers have so far revealed the following about yawning.[2]

Yawning is a common and probably universal human behavior. Other animals yawn, but the significance of this behavior in creatures from fish to lions is not clearly understood.[3]

In humans, yawning is performed throughout life. Dr. Richard Roberts of the Genetics and Prenatal Diagnostic Center in Signal Mountain, Tennessee, studied ultrasound scans of newborns and found that the fetus yawns (and also hiccups) as early as eleven weeks of age.[4]

According to Dr. Provine, a yawn technically involves a "gaping of the mouth accompanied by a long inspiration followed by a shorter expiration." Yawning is important in opening the eustachian tubes (which run from the ears to the throat) and in adjusting the air pressure in the middle ear.

Yawning is of clinical importance in health. Yawning, or its absence, can be a symptom of migraine, brain lesions, tumors, hemorrhage, motion sickness, chorea, encephalitis, or Parkinson's disease. It is also an important therapeutic factor in preventing postoperative respiratory complications. It has been reported that psychotics rarely yawn, except in the case of brain damage. Some clinicians claim that those with acute physical illnesses do not yawn until they are on the road to recovery.

Yawning is commonly associated with drowsiness, boredom, and low levels of arousal. Studies confirm, for example, that humans are more likely to yawn when participating in lengthy, uninteresting, or repetitive tasks, and yawn more when observing uninteresting rather than interesting phenomena.[7]

THE AVERAGE life span of a human taste bud is seven to ten days.

STUDIES SHOW that 25 percent of people who lose their sense of smell also lose their sex drive.[6]

NOSE-PICKING in public is actually encouraged in Macau.

Dr. Provine and colleagues believe that yawning is grossly underresearched. "Beliefs about the relation between sleepiness and yawning," they add, "are based upon folk wisdom and everyday observations to which science has had little to add." For instance, texts on sleep only occasionally mention the prominence of yawning in drowsy people but typically cite only unrelated or general references on yawning.

Neurological evidence for an association between yawning and stretching comes from case reports of brain-damaged individuals who are unable to separate the two behaviors. During yawns, such persons often perform related stretching movements of otherwise paralyzed body parts. Studies show that drugs that produce yawning also produce stretching in a variety of animals.[8]

Dr. Provine's studies show that there is also "evidence for at least partial autonomy" of yawning and stretching, and one of his laboratory experiments showed that "whereas 47 percent of stretches were accompanied by a yawn, only 11 percent of yawns were accompanied by a stretch."

Furthermore, he theorizes that yawning shortly before sleeping and after waking may be either a mechanism to increase alertness or brain function in a drowsy person, or a mechanism to depress alertness, encourage relaxation, or hasten or otherwise prepare us for sleep.

Dr. Provine and his colleagues note that while "only a few hypotheses about yawn function have been evaluated," there is no support for the popular assumptions that yawning is either a response to or somehow regulates blood levels of carbon dioxide or oxygen. The yawning rate, they found, is neither facilitated nor depressed by breathing gases with elevated levels of carbon dioxide or oxygen. The researchers report that yawning is also unaffected by vigorous exercise.

We also know that yawners who yawn infrequently do not compensate by performing

THE HUMAN EAR can detect sound frequencies between 20 and 20,000 hertz.

EARRINGS ARE worn for many reasons. Sailors first started wearing gold earrings so they could afford a decent burial when they died.

THE AVERAGE talker sprays about three hundred microscopic saliva droplets a minute. This is about two and a half droplets per word.

yawns of longer duration, nor do frequent yawners perform shorter yawns.

One more thing. If Dr. Provine and colleagues are correct in all points mentioned above, you have probably yawned at least once while reading the last several paragraphs.[9]

Why do my eyes and mouth
sometimes water when I yawn?

Watering eyes may be the result of pressure on the main tear glands (located at the outer margins of the eye sockets) caused by the facial contortions involved in yawning. The involuntary act of yawning usually includes opening the mouth very wide while slowly taking in a deep breath.

These same contortions might also put pressure on the salivary glands, especially in a stifled yawn, in which the yawner struggles to keep the mouth closed while opening the throat widely. The three pairs of salivary glands are located over the angle of the jaw, in the floor of the front of the mouth, and toward the back of the mouth close to the sides of the jaw.[10]

Why do people hear my
voice differently than I do?

This is a classic OBQ. People have been wondering about this "body oddy" since tape recorders were first invented. When you hear

EVERY TONGUE print is different.

THE TONGUE IS the strongest muscle in the human body.

A PERSON suffering from excessive thirst is experiencing polydipsia.

your voice played back on a tape recorder for the first time, you can't believe your ears. "I don't sound like *that* do I?" No one can convince you that you do.

Actually, in one sense you're right. It has been discovered that questioning, disputing, or completely rejecting a recorded voice as one's own is a worldwide occurrence. Anthropologists have been reporting this for years. They are often the first to introduce tape recorders to the diverse cultures throughout the world as a tool used in their linguistic research.

In fact, there is a simple explanation for this curious and seemingly universal auditory phe-nomenon. According to Dr. Nelson Vaughan, a retired speech therapist, diction instructor, and voice coach in Hollywood, when we listen to our own voice while we speak, we are not hear-ing solely with our ears. We are also "internally hearing a mostly liquid transmission through a series of bodily organs."

Speech begins at the larynx, from which a sound vibration emanates. Part of this vibration is conducted through the air. This is the part that *others* hear when we speak, and what tape-recorders record. But another part of the vibration is directed through the various fluids and solids of our head. Our inner and middle ears are located within caverns hollowed out of bone—in fact, the hardest portion of the human skull. The inner ear contains fluid, and the middle ear contains air. Both constantly press against each other. The larynx is also encased in soft tissue full of liquid.

Sound transmits differently through air than through solids and liquids. This difference accounts for nearly all the tonal variations we hear compared with what others hear.

Dr. Vaughan points out that even young children experience this auditory curiosity. Often they will completely fail to recognize their own voice when it is played back to them, even if the playback occurs immediately after they spoke.

Research fails to show that the voice we hear (our internal voice) is either necessarily pitched higher or lower than the voice others hear (our external voice). Therefore, it can be said that we all really have two voices. Either one is equally our "real" voice.

It's all in the ear of the beholder.[11]

Why do my ears ring after hearing a loud noise?

This common phenomenon is actually a harmless medical condition known as temporary tinnitus. But chronic tinnitus is something else again. It can have serious health consequences, and in extreme cases chronic tinnitus sufferers have been known to commit suicide in order to escape the maddening "ringing."

As for temporary tinnitus, sound normally comes from a source outside our own body. The sound stimulates the ear's auditory nerve, and the brain interprets these impulses as noise.

ARACHIBUTYRO-phobia is the fear of peanut butter sticking to the roof of your mouth.

THE AVERAGE person ingests about one ton of food and drink each year.

Through experience, we come to distinguish among types of noises and to judge which are significant and which are insignificant. Strangely, a person with tinnitus receives auditory nerve stimulation from something other than an external source.

Theoretically, virtually anything that can disturb the auditory nerve can cause chronic tinnitus, including extremely loud noises. For example, when a cannon is fired right next to where one is standing, the auditory nerve may remain irritated for several seconds after the shot is fired and the sound ceases. Nevertheless, the irritated auditory nerve still sends impulses to the brain and those impulses are interpreted by the brain wrongly as noise. Hence, we experience the ringing until the auditory nerve is no longer irritated. In this case, the tinnitus is only temporary.[12]

According to Dr. Tuan Pham, an ear, nose, and throat specialist at the Royal Prince Alfred Hospital in Sydney, chronic tinnitus "needs to be further investigated." Dr. Pham points out that, besides the ringing sensation, chronic tinnitus sufferers can hear "hissing," "buzzing," "humming," "a waterfall-like sound," or "a pulsatile sound" (corresponding to one's own heartbeat).

Besides being a reaction to a loud noise, he adds, temporary tinnitus is most commonly caused by vascular distress after a physical or mental trauma, or by an allergic response to

DOGS, PIGS, AND some other mammals can taste water. Humans cannot. Humans do not actually taste the water; they taste the chemicals and impurities in the water.

IF YOU WANT TO be an astronaut, you'd better have absolutely no sense of smell. In a space vehicle, you can't open a window to let stale air out.[13]

medication. Fluid tablets or aspirin may be involved as well. In one study he cites, people who took twenty aspirins a day and were subject to tinnitus attacks afterward saw these attacks completely disappear when they discontinued the large aspirin doses.

The causes of chronic tinnitus are numerous. These include inflammation of the ear, ear canal problems, jaw problems, work noise exposure, old age, clogging of the ear by earwax, vertigo attacks, excessive use of the telephone, muscle spasms in the ear, nutritional deficiencies, and allergies and infections, among other factors.[14]

Chronic tinnitus sufferers not only have to live with the infernal noises, they often experience hearing loss as well.[15]

According to Dr. Pham, there are various measures to help a patient cope with chronic tinnitus. Chief among these is "reassurance."[16]

ON AVERAGE, females have shorter vocal chords than males.

THE AVERAGE smell weighs 760 nanograms. A nanogram is one-billionth of a gram. Japanese researchers weighed odors by dissolving them in fat and using an ultrasensitive quartz crystal.

Can I permanently change the sound of my voice?

You can definitely change the sound of your voice. A speech pathologist claims that 30 percent of people are unhappy with the sound of their voice. Dr. Daniel R. Boone of the Department of Speech and Hearing Science at the University of Arizona at Tucson, adds that "older people want to sound younger, and the young want to sound older." And with more business being conducted over the

YOU'LL DRINK about twenty thousand gallons of water in your lifetime.

THE AVERAGE person laughs about fifteen times a day.

WHEN YOU laugh, you expel short bursts of air up to seventy miles an hour.

ON AVERAGE, people can hold their breath for one minute. The world record is seven and a half minutes.

A SNEEZE CAN exceed a speed of one hundred miles an hour.

phone, your "voice is becoming as important as your looks."

According to Dr. Morton Cooper, a Los Angeles speech pathologist, "we live in a society where we want to sound a certain way. We want our voice to sound not nasal but rich and full. We want authority in our voice. We want to sound as if we have status and position."[17]

Dr. Lillian Glass, a speech pathologist in Beverly Hills, California, is known as "speech consultant to the stars." Her clients have included Dustin Hoffman and Sean Connery. She claims: "Often you can give a false impression by the way you speak. A rough, harsh voice can give the impression that a person is mean and gruff. A high-pitched sound can signal immaturity. If you have a monotonous voice, people may tend to think you're not interesting—or interested."[18]

Just like cosmetic surgery, voice improvement is becoming more and more fashionable. Professional speech consultants in the United States are doing a booming business. It seems that any voice can be improved. Speed, tone, degree of nasality, even accent, can be changed. According to Dr. Boone, "sometimes all it takes is just training people to speak a little faster." In the United States, where the southern accent is seen as "lower class," many individuals from the

South seek to obtain a different accent. Dr. Lawrence Stewart, emeritus professor of educational psychologist at the University of California at Berkeley, says, "When people hear a southern drawl, they automatically deduct twenty IQ points."

Dr. Glass adds that six months of voice training can work wonders, because "nobody is born with a bad voice. Anybody can improve the way they speak."[19, 20]

Why is the Australian accent more pliable and the U.S. accent more durable?

Linguists have fun with this one. It has long been noted that when Australians live in the United States for six months or more, their Australian accent begins to become "Americanized." Eventually, they sound just like Americans. Listen to how Mel Gibson, Olivia Newton-John, and Nicole Kidman sound now compared with how they sounded when they left "the land of Aus." But when Americans live in Australia for six *years* or even much longer, their American accents do not change at all.

Why? It seems that no one can account for this.

When Australians sing American songs, why can't we hear an Australian accent?

According to the experts, you can in fact detect some traces of accent, depending on the singer and the lyrics of the song. Dr. Charles Diggs, a speech-language pathologist with the American

Speech-Language-Hearing Association in Rockville, Maryland, says that many people show such accents but that there may be psychological and physical reasons for sounding "American-like" in an American song.

Many singers from a particular national background, Dr. Diggs notes, do a lot of what is called "code switching" when verbally communicating with those of another background: "They may subconsciously choose a different language and rhythm to make themselves more clearly understood, much as an adult often uses slower, simpler speech when talking to a child. Thus, a singer might subconsciously or consciously choose [to sound American]—thinking that's what the audience wants to hear." Dr. Diggs concludes, "With singing, you already have words and a rhythm and rate of speech provided. If you already have the words, you can concentrate more on the manner in which you are saying them."[21]

Can illness cause a change in accent?

Sometimes an accent can completely baffle the experts. A U.S. doctor has reported the strange case of a thirty-two-year-old Baltimore man who suddenly began speaking with a Scandinavian accent. The man displayed a rare disorder that may shed light not only on how the brain recovers from a stroke but also on how the brain produces language.

The man suffered from "Foreign Accent Syndrome." As a monolingual American-English speaker, he had no previous experience whatsoever with any foreign language. Nothing in his past pointed to fluency or even familiarity with any foreign language.

Yet when he spoke after suffering a stroke, he sounded both Nordic and totally unfamiliar with English. Dr. Dean Tippett, a neurophysiologist at the University of Maryland School of Medicine in Baltimore, adds: "He was pretty clear. Everyone who heard him said he sounded Scandinavian or Nordic."

Foreign Accent Syndrome is a neurological condition in which a brain malfunction produces speech alterations that sound like a foreign accent. The syndrome is triggered by stroke or head trauma. There are reports that Foreign Accent Syndrome has produced German, Spanish, Welsh, Scottish, Irish, and Italian accents.[22] So far, no case of an Australian or American accent developing has been reported in the literature.[23, 24]

In a 1990 paper presented at an American Neurological Association meeting in Atlanta, Dr. Tippett argued that studying the intricacies of Foreign Accent Syndrome may reveal secrets about how particular parts of the brain contribute to spoken language. Monitoring the course of the return of a patient's original accent may be a useful means of tracing the process of stroke recovery as well.

In the case of this patient, immediately after the stroke the man's speech was slurred for a day or two. His Scandinavian accent appeared only

OUR TWO nostrils register smell in different ways. The right nostril detects the more pleasant odors. The left nostril is the more accurate.

THE NOSE IS our personal air-conditioner. It warms cool air, cools warm air, and filters impurities.

IN ANCIENT Rome, to be born with a crooked nose was considered a sign of being a good leader.

THE TECHNICAL name for a habitual nose-picker is rhino-tillexomaniac.

BY AGE SIXTY, most people have lost half their taste buds.

as he started to recover. He added extra vowel sounds as he spoke, saying such things as, "How are you today-ah?" His voice also rose in pitch at the end of sentences, as if asking a question. Moreover, "that" was pronounced as "dat." Some vowel sounds were substituted as well, making "hill" come out as "heel" and "quite" come out as "quiet." Often the vowel sound was exaggerated and drawn out.

Perhaps the world's foremost authority on Foreign Accent Syndrome is Dr. Arnold Aronson, a speech pathologist at the Mayo Clinic in Minneapolis who has evaluated about twenty people with the syndrome. According to Dr. Aronson, cases involving non-Americans have produced a French accent in a British person and a Polish accent in a young Czech. It is interesting that about 40 percent of cases produced German, Swedish, or Norwegian accents.

According to another expert on the subject, Dr. Elliott Ross, the trauma-acquired foreign accent may become "rather permanent," depending on where the brain injury is. Dr. Ross is the director of the clinical research program at the Neuropsychiatric Research Institute in Fargo, North Dakota. In the case of Dr. Tippett's patient, the man's speech returned to normal about four months after the stroke.

Dr. Tippett reported that his patient suffered no additional pain in speaking differently. Indeed, at first the man even "enjoyed his new accent, saying he hoped it would help attract women." But by the time his accent had almost completely faded, he said he was happy to be speaking like an American again.[25, 26]

What causes "varicose nose"?

What do legendary comedian W. C. Fields, Santa's reindeer Rudolph, and thousands of people throughout Australia have in common? Their bright red noses, of course. National Red Nose Day is an annual event in Australia. Many of the famous and not-so-famous will pay for the privilege of wearing a plastic red nose in support of Sudden Infant Death Syndrome research, counseling, and support services.

But what of the red nose condition itself? What is it called? What causes it? How can it be prevented or treated?

Contrary to popular opinion, the red nose condition is not called "varicose nose." Nor is it caused by too much alcohol, although alcohol can aggravate the problem. W. C. Fields was a Hollywood star who was known to drink heavily. His alcoholism and "varicose nose" came to be associated, but there is no causal connection. Instead, the condition is actually a form of acne called rosacea.[27]

In the *Journal of the American Medical Association,* Marsha Goldsmith explores developments in the treatment of rosacea.[28] For years, doctors have treated this condition with antibiotics, often without success, but a few years ago doctors at Harvard University found that a

YOU CAN'T catch a cold by sitting in a draft. The change in temperature may bring on a temporary condition called vasometer rhinitis. This causes swelling in the tiny blood vessels in the mucus membrane linings of your nose and gives you a runny nose, but it is not a cold. A cold is a virus.

109

cream commonly used to treat (of all things) vaginal infections is much more effective.

The doctors tested the cream, which contains metronidazole, on forty men and women with moderate to severe rosacea. After three weeks, the cream had cleared up half the lesions and redness and relieved the itching and dryness that accompany rosacea. Among those who used a placebo cream instead, there were no noticeable changes. The doctors are not sure why the cream works, but they think that perhaps metronidazole kills a mite called *Demodex folliculorum* that burrows into the pores of the skin. The presence of too many of these mites can inflame the skin.

Dr. Jonathan Wilkin of the Department of Dermatology at Ohio State University says: "Rosacea is truly a cutaneous disorder, not just 'a complexion problem that runs in my family,' as many people think. It can be treated easily now and reversed. If not treated, it will probably grow worse."

Every year on National Red Nose Day, Australians will be putting on red noses for one day for a good cause. But for others, they may one day remove their red noses for good.[29]

What is a hiccup?

The medical name for a hiccup is singultus— hence, don't sing alone: sing ult us!

A hiccup is actually an irritation of the diaphragm that causes a spasm. In fact, a hiccup is a two-part phenomenon. First, the diaphragm contracts involuntarily because the nerves that control it have become irritated by something. Perhaps we ate or drank something too fast. When breathing and eating have to be undertaken at the same time, you invite irritation. Usually you control your own hiccuping by not doing things that provoke hiccups.

Second, when air is inhaled the space between the vocal cords at the back of the throat (the glottis) snaps shut with a characteristic clicking sound. This snapping is what we hear when we hiccup.

To understand a hiccup, you have to understand what the diaphragm is. The lungs are enclosed in a kind of covered cage in which the ribs form the walls and the diaphragm forms the floor. The diaphragm is an upwardly arching sheet of muscle. When you take a normal breath, the diaphragm is drawn downward until it becomes flat. At the same time, the muscles that surround the ribs contract, lifting the lungs up. It is rather like the lifting of a hoop skirt in an updraft. In this way, the chest cavity becomes wider and deeper, and its air capacity increases.

There are many remedies for hiccups. Some people drink a glass of water without pausing for air. Others hold their breath until the hiccups stop. Still others breathe into a bag. These techniques may

UNTIL BABIES are six months old, they can breath and swallow at the same time, but adults cannot.

THE SOUND OF A snore can be as loud as a pneumatic drill.

WE HAVE FOUR basic tastes. The salt and sweet taste buds are at the tip of the tongue, the bitter taste buds are at the base, and the sour taste buds are along the sides.

IT REQUIRES THE
use of seventy-
two muscles to
speak a single
word.

A ONE-MINUTE
kiss burns
twenty-six
calories.

IF IT WAS POSSIBLE
for the human
voice to be
carried naturally
for great
distances
through the air, it
would take four-
teen hours for a
shout bellowed
in Australia to
be heard in
California.

restore the normal rhythm of the twitching, irri-tated diaphragm. Perhaps this happens by reduc-ing the oxygen level and increasing the carbon dioxide level. Other cures attempt to trick the nervous system with such diversionary tactics as tickling the nose to induce sneezing, or pulling the tongue. If hiccups persist in spite of all efforts to stop them, medical attention should be sought.

Sometimes hiccuping can be caused by a dis-ease. For example, there is a medical condition called epidemic hiccuping. It is a symptom of some forms of encephalitis (inflammation of the brain).

The record for the longest span of uninter-rupted hiccuping is (are you ready for this?) sixty-eight years. The unfortunate man hiccuped an average of twenty to twenty-five times a minute—but otherwise he led a normal life, mar-rying twice, and fathering eight children.[30]

What is earwax?

It's sticky, it's ugly, it's yucky, but it's important to health. It's earwax.

The medical name for earwax is cerumen. Earwax is sticky on pur-pose. Dust, dirt, bacteria, fungi, and other foreign dangers to the body all stick to the wax and thus do not enter the ear—one of the most sen-sitive areas of the body, and quite exposed when you think about it.

Earwax also contains special enzymes called lysozymes, which break down the cell walls of foreign bacteria and are also contained in saliva. So earwax fights bacteria in two ways—sort of acting like flypaper to halt them and then biochemically dissolving them.

Also, there are different colors of earwax among the various peoples of the world. In white and black people, earwax is honey-colored, moist, and soft. But in certain Asian groups (Mongolians, for instance), it is gray, dry, and brittle. There is a specific gene for earwax. Wet is the dominant trait, dry is recessive.

Although we are taught to clean our ears of earwax, it is probably a good idea, from a health standpoint, to leave some of it there. Just as it's wise to be lucky, it's wise to be yucky![31]

How do we swallow?

A normal swallow is a three-phase process taking from eight to twelve seconds from beginning to end. Part of the swallow is voluntary, but most is involuntary.

AS FAR BACK AS the sixth century, it was customary to congratulate people who sneezed because it was thought that they were expelling evil from their bodies. During the great plague of Europe, the pope at the time made it law to say "God bless you" to anyone who sneezed.

The first phase of food swallowing is the termination of chewing. Chewing is the first and last voluntary aspect of the swallow. Chewed food in the mouth is mixed with saliva to prepare it for passage down the throat (pharynx). The actual swallow begins as the food is pushed to the back of the throat by the tongue. The tongue does this by

ABOUT FIVE
thousand
different
languages are
spoken on Earth.

THE AVERAGE
person spends
about two
weeks of their
life kissing.

APPROXIMATELY
forty million
Americans
suffer from
chronic bad
breath
(halitosis).

THE AVERAGE
yawn lasts for
six seconds.

A FETUS MAY
begin yawning as
early as eleven
weeks after
conception.

pressing against the roof of the mouth (soft palate), forcing the food into the throat.

The second phase is when the food is in the throat. This takes about two seconds and is involuntary. These involuntary reflexes are controlled by the swallowing center located in the brain stem. Food is pushed down into the esophagus by rhythmical contractions of the esophageal muscles (peristaltic waves). It is important that the sphincter muscle at the entrance to the esophagus remains relaxed in order to open the channel of the throat. Also, the larynx must elevate to force the epiglottis to close over the airway (trachea) and thus prevent food from entering the lungs.

In the final phase, the peristaltic waves continue to push the food through the esophagus and down into the stomach. One more muscle sphincter must relax for this occur, and also to help prevent food from coming up again. This last phase takes between six to ten seconds.

With such a complicated process, it is no wonder that something can easily go wrong.[32]

Does my heart stop when I sneeze?

This is one of the most commonly held body myths. Your heart does not stop when you sneeze, although it might feel as if it does. Instead, when we sneeze a powerful positive pressure condition is created in the chest. It is fairly violent and can actually change the rhythm of the heartbeat. But it does not stop the heart from beating.

This feeling of a momentary jump in the heartbeat is probably the origin of the erroneous belief that the heart stops. According to Dr. Jay Block, former president of the American College of Chest Physicians in New York, the creation of positive pressure in the chest when we sneeze or cough has a name—the Valsalva maneuver.[33]

Why do I sometimes sneeze when I come out of a dark room into the daylight?

Even if this doesn't happen to you, just watch people leaving a theater at the end of a matinee to see that it certainly happens to others.

Francis Bacon (1561–1626) discussed this "light sneezing" in his *Sylva Sylvarum* (1635). He erroneously believed that it was caused by moisture being drawn down from the brain to the nostrils and eyes due to a "motion of consent."[34]

But consent has nothing to do with it. According to Dr. R. Eccles of the Common Cold and Nasal Research Centre of Cardiff, Wales, this is termed the "photic sneeze." In fact, it is a

THE AVERAGE mouth produces 1.8 pints of saliva each day.

YOU'LL FEEL thirsty if the amount of water in your body is reduced by just one percent.

IF YOU TEND TO chew your food on the right side of your mouth, you're probably right-handed. If you tend to chew your food on the left side of your mouth, you're probably left-handed.

genetically transmitted characteristic that affects between 18 and 35 percent of the population. Dr. Eccles writes: "The sneeze occurs because the protective reflexes of the eyes (in this case on encountering bright light) and nose are closely linked. Likewise, when we sneeze, our eyes close and also water. The photic sneeze is well known as a hazard to pilots of combat planes, especially when they turn towards the sun or are exposed to flares from anti-aircraft fire at night."[35]

Does an earlobe crease indicate high risk for heart disease?

In the early 1970s it was found that the presence of an earlobe crease (a line as if an earring had been torn through the lobe), especially a vertical crease, was correlated with a somewhat higher incidence of heart disease. Over the years, however, studies have disputed the correlation and also failed to establish the mechanisms by which an earlobe crease could be linked to heart disease.

It has been suggested that a decrease in blood flow might result in the collapse of tissue and cause a crease, because the earlobe has a rich blood supply, but this notion has been largely dismissed. It has also been observed that both earlobe creases and heart disease occur more often in the elderly—and we see more of one because we see more of the other these days.[36]

Does the ability to smell really well run in families?

There is some evidence to support this view. It has been discovered that the rough surfaces of our nasal cavities play a crucial role in our ability to smell. Deeper and more intricate cavities can allow smell molecules to remain in the nose longer and hence be better perceived. The pattern of nasal cavities is genetically determined and runs in families.[37]

chapter 6
the **skin**

What is the largest human organ? You were right if you said the skin. The skin is also the heaviest organ, weighing between about five and a half and ten pounds. If you were to spread it out like a pancake, your skin would cover an area of about six square feet. It forms a perfect protection for our body, provides great insulation

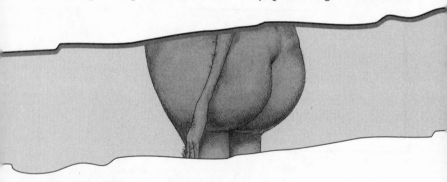

against heat and cold, and keeps out bacteria and other things that would harm us. It also serves as the end point for our nerves so that we can touch. In fact, skin is pretty fantastic!

Why do I itch?

Although science tells us much about itching, we still don't understand everything about what is medically termed "pruritis."

Sensory receptors located just below the surface of the skin send messages to the brain. Itch sensations flow along the same pathways of the nervous system as pain sensations. The vast majority of sensory receptors are "free" nerve terminals that do not seem to be designed for any single function. They carry both itching and pain messages.

These sensory receptors are the most common nerve terminals in the human body's entire nervous system. This is understandable, because the entire surface of the body must be covered. When these sensory receptors operate at a high level, a pain signal results—but an itch signal results when only a low level registers.

Scientists can induce itching by heating a subject's skin, but if too much heat is applied, pain is produced instead. Certain chemicals can also bring about itching in humans during laboratory experiments. Chief among these are the histamines. This is why doctors often treat pruritis complaints with antihistamines.

YOUR fingerprints normally do not change over the course of your life. However, scarring can change fingerprints, as can hard work with your hands that can eventually wear down some of the ridges of your fingerprints.

For example, when a doctor has a patient who complains of itchy hives, the doctor treats the hives and the itchiness stops. But this does not mean that anyone knows for sure why the itch is associated with the hives. Itching can also be associated with a variety of serious illnesses (such as Hodgkin's disease) and can indicate the onset of others (such as diabetes).

The function of itching is open to speculation. Some physiologists maintain that itching provides a first warning to the body of imminent pain. Some anthropologists suggest that in our primitive past itching served the purpose of telling early humans when it was time to pluck out body lice and other skin parasites.

Actually, itching may be less easy to live with than pain. Doctors note that patients with severe itching are invariably more willing to scratch till they bleed.

Itching has been compared to tickling, but evidence suggests that these reactions are entirely separate. Ironically, although tickling produces laughing, while itching and pain do not, subjects can take only short periods of tickling compared with the other two. In this sense, it is not always true that one chooses pleasure over pain.

New sources of itching appear continually. According to Dr. S. Leonard Syme of the School of Public Health at the University of California at Berkeley, there seems to be an epidemic of *pruritis ani*—anal itching due to overvigorous wiping with abrasive toilet paper.[1]

Nevertheless, science has much to learn about itching. Dermatology professor Dr. Ronald Marks of the College of Medicine of the University of Wales remarked: "There are many areas with persistent pruritis that we find enormously difficult to handle clinically. Treatment with conventional therapies is hopelessly inadequate."

One myth about itching has definitely been dispelled. There is no evidence for the so-called "seven-year itch." However, George Axelrod's famous play and subsequent classic Billy Wilder film by that name—about the roving male libido after seven years of marriage—is a cherished part of our folklore, even if not our medicine.

There are many mysteries of the "Itchy and Scratchy Show" (of "The Simpsons" fame) taking place in our bodies. Dr. Gil Yosipovitch, a professor of dermatology at the Baptist Medical

Center of Wake Forest University in Winston-Salem, North Carolina, says, "No one has figured out why it is so pleasurable to scratch an itch."[2]

The enigma of itching is irritating—and scientists continue to scratch.[3]

YOUR BELLY button too increases and decreases in size as you gain or lose weight.

Why do I get chills when I hear chalk screech on a blackboard?

It crawls up and down your spine. You wince, you shiver, you cringe at the mere thought of it. Your worst enemy or perhaps your best friend might have done it to you. The ancient Greek philosopher Aristotle first observed it and remarked on its chilling consequences. Modern lexicographers have resurrected the archaic verb "gride" to describe the act of causing it. And science is still puzzled by it.

What is it? It is that awful, high-pitched, scraping sound you hear when chalk is dragged across a blackboard—"blackboard screech."

Why do we get chills when we hear blackboard screech? The short answer is that we don't really know. Despite the antiquity of its discovery and the near universality of its spine-tingling effects, surprisingly little is actually known about this odd, auditory phenomenon.

No scientific studies examined blackboard screech until 1986. Previously, it was commonly believed that the high-frequency sounds were responsible for the unpleasantness experienced by those hearing it. However, when this theory was tested in the laboratory, it was proven to be incorrect.

In 1986, three U.S. researchers at the Cresap Neuroscience Laboratory of Northwestern University conducted four experiments to test the psychoacoustics of blackboard screech.[4] But these experiments actually raised more questions than they answered.

The researchers, Drs. D. Lynn Halpern, Randolph Blake, and James Hillenbrand, first re-created blackboard screech by dragging a garden rake across a slate surface. They recorded the sound and ran the recording through different frequency processors.

ALTHOUGH THE skin is the largest body organ by weight, the surface area of the skin is vastly smaller than the surface area of the lungs. The lungs are made of millions of small grape cluster–like groups called alveoli. This is where oxygen exchange between the air and the blood takes place.

The original theory that blackboard screech produces chills simply because of its high frequency quickly fell by the wayside. When they removed the high frequencies, it had no effect on the unpleasant reactions experienced. In fact, when the lower frequencies were removed, the sound that was left was judged by subjects in the experiments to be rather pleasing. Next, the researchers experimented with the volume, but volume had no bearing on the chilling effect.

Finally, they compared blackboard screech with a variety of sounds found in nature and discovered that blackboard screech bears a remarkable resemblance to the warning cries emitted by a species of lower-order primates: Japanese macaques. This result suggested to the researchers that our tingling spines may be a primitive reflex left over from an earlier phase of our evolutionary history. It is also possible that blackboard screech may be similar to the sound made by some pre-historic predator. Yet these are only conjectures.

The researchers concluded: "Regardless of this auditory event's original functional significance, the human brain obviously still registers a strong vestigial response to this chilling sound."[5]

Why does my skin wrinkle?

The conventional answer is that wrinkles are caused mainly by exposure to the sun's ultraviolet radiation. Over time, the ultraviolet rays break down the middle layer of the skin (the dermis), causing it to loosen and allowing wrinkles to form. This condition is known as "dermatoheliosis"—sun-induced skin aging. Yet, with extreme old age, the dermis loosens of its own accord.

According to Dr. Allen Lawrence, a noted dermatologist in Chicago, wrinkles are more likely to occur among people who have thin layers of skin or whose skin is lighter in color.[6] Exposure to the sun is to be avoided as much as possible to preserve the skin. The next best step is to use a sunscreen.

Aged skin is thinner than younger skin, and less cellularly organized. Under the microscope, one can see in younger skin the healthy basal cells lined up in neat columns within the outer layer of the skin (the epidermis), but magnification shows that aged skin is in disarray from a breakdown in the normal process of cellular proliferation and organization.

With age, collagen fibers, which are important in maintaining skin integrity, decrease in number, organization, and density. Also with age, the formerly smooth, ribbonlike elastin fibers become

ODD-LOOKING, blue-skinned people live in Troublesome Creek, Kentucky. For more than 160 years, some of these individuals, all believed to be related to one immigrant from France, have had blue skin caused by a defective gene. Their bodies lack an enzyme needed to convert a blue protein in the blood into the red protein hemoglobin. That gives them their bluish color.

coarser, denser, and less resilient. Elastin fibers are responsible for the skin's ability to snap back to shape after being stretched, particularly after the skin has been exposed to sunlight.

Other things happen when skin ages. The tiny blood vessels in the dermis become thick-walled, yet also leakier. There is a general loss of nerve cells, hair, and sweat ducts, and the sebaceous glands, which produce fatty secretions called sebum.

Skin aging is a combination of two processes, one intrinsic, the other extrinsic. Chronological aging is genetically programmed, but the chemical reactions triggered by exposure to sunlight are environmental. We may not be able to do anything about our genes, but we certainly can control our behavior and stay out of the sun.

The medical term for a wrinkle is (are you ready for this?) "wrinkle." The origin of this word is obscure, but it probably comes from the Old English word "gewrinclod," which means "the winding of a ditch."[7]

Why do the ends of my fingers and toes wrinkle after a bath?

This is one of the classic OBQs. The short answer is that this form of wrinkling occurs because of differences in the layers of skin and our skin's natural protective oils.

According to Dr. Marianne O'Donoghue, a professor of dermatology at Rush University in Chicago, when we are immersed in a bath "the top layer of the skin absorbs a lot of water. The bottom layer of skin can't get any bigger, so the top layer must corrugate or pleat. Fortunately, the effect is reversible."

But, you might ask, why do we shrivel up like a piece of dried fruit and not swell up like a sponge? Without our skin's protective oils, the palms of our hands and the soles of our feet become *dehydrated* in the presence of water, not *hydrated*.

It works this way. About 75 percent of our body is water, the exact amount depending on the amount of fat we have. Dehydration occurs when the protective oil of our skin surface is washed away. At that point, our body water begins to leak out of the skin cells. These cells have semipermeable membranes, which means they can easily give up water but cannot absorb it as well. Once the oil is lost, after about fifteen minutes of immersion, the floodgates open outward. The water moves out and wrinkling occurs.

The ends of our fingers and the palms of our hands wrinkle more quickly than the backs of our hands, because the backs of our hands have extra sebaceous glands. These glands continually replenish those protective oils almost as fast as they get washed away.[8]

Is it possible to stop wrinkles?

This book is trying to stay away from therapeutic suggestions, but we just can't help ourselves here. Are those first tiny crow's feet starting to appear at the outside corners of your eyes? And what about under them? Are those forehead lines really caused by smiling? How would you really like to laugh at your first wrinkles—and at all those yet to come?

A U.S. medical team has discovered a new elixir that stimulates growth of skin cells. Applied as a cream, it could well prove to be the fountain of youth for aging skin. The elixir is composed of the hormone PTH minus a few amino acids. PTH is produced in certain neck glands. When Dr. Michael Holick and colleagues at the Section of Endocrinology, Diabetes, and Metabolism at the Boston University School of Medicine injected laboratory rats with the PTH compound, there was a dramatic 300 percent increase in the rate of skin cell growth. The rats emerged with younger, softer skin.[9]

One reason that skin looks wrinkly with age is that skin cells are not replaced as fast. The PTH compound may also be useful in treating scarring, burns, and even baldness.

The experiments with humans are under way. Hopefully, what will work in rats will work in humans. After all, Dr. Holick agrees, "the two are pretty similar in lots of other ways."

What are stretch marks?

According to Dr. Alan Xenakis in *Why Doesn't My Funny Bone Make Me Laugh?* (1993), stretch marks are merely evidence that the skin has been forced to expand in order to accommodate a somewhat larger body mass. The skin has an amazing ability to expand and contract as the body mass changes. Stretch marks often occur after dieting or a pregnancy.[10]

It is a myth that stretch marks occur more often in women than in men, although it may seem so. One reason for this is that hair is more likely to cover men's bodies in places where stretch marks are most likely to be seen.

Where are those places? You're right about that too.[11]

What are warts?

Warts are infectious swellings or benign tumors in the outer layer of the skin. Because they are contagious, they can spread from person to person. Most frequently, they appear on the hands, fingers, and soles of the feet. However, genital warts can occur on the genital and anal areas.

Common warts are caused by the human papilloma virus, which can remain inactive in the body as long as six months before warts appear.

The medical term for a wart is "verruca." There are at least twenty-six types of verruca. The common wart is sometimes called the "seed verruca," because of the tiny black specks or "seeds" within it. In fact, these "seeds" are actually elongated capillary loops that have thrombosed (clotted). Another interesting verruca is the "soot verruca." This is the so-called "chimney sweep cancer" that was common among chimney sweeps in such cities as London in the nineteenth century. But to call this wart a benign tumor is a misnomer. "Chimney sweep cancer" is really a malignant carcinoma of the scrotum that is the result of poisoning from chimney soot.

It was one of author Roald Dahl's jokes that he named one of his characters after warts. This is the obnoxious Veruca Salt ("the girl who got everything she wanted") in *Charlie and the Chocolate Factory* (1964). Why Dahl dropped an "r" in the spelling of her first name is known only to him. His subtle humor also extended to Veruca's surname. Throwing salt on a wart is one of the traditional folk customs for getting rid of them.

As people age, they tend to build up immunity to warts. Nevertheless, warts can occur at any age.

Studies show that two out of three cases of warts disappear on their own within two years, which is why doctors now usually recommend leaving warts alone.

But there are two conflicting theories about the treatment of warts. One view says that, because warts occur most commonly in children, children should not have warts removed, because having to cope with the virus can boost the immune system. The other view holds that only *some* of the warts should be removed, because if one or two warts are treated, others on the same person disappear without treatment. Removing the warts may stimulate antibodies to fight the virus.

Normally, warts do not themselves endanger health, but you should see your doctor about any wart you might have—just to be sure. Genital warts should also be seen by a doctor, for there is evidence that they lead to cervical cancer. Plantar warts (on the soles of the feet) are particularly painful and also need a doctor's attention.

THE SKIN OF THE armpits can harbor up to a whopping 516,000 bacteria per square inch, while drier areas, such as the forearm, have only about 13,000 bacteria per square inch.

Treatment for warts varies, Over-the-counter treatments often involve salicylic acid. Cryotherapy (freezing) is often employed, as is also, sometimes, surgery, including laser surgery. In 1994, a story in *Australian Doctor* described how the drug cimetidine was "remarkably effective at curing warts in children."

Hypnotherapy has also been used to treat warts. According to Dr. Robert Noll of the Children's Hospital Medical Center in Cincinnati, it seems to be particularly effective in treating children. Dr. Noll writes, "Nearly all patients (86 percent) were completely cured of their warts within three months of the onset of therapy."[12]

The fact that most warts vanish by themselves after two years increases their mystery. It also accounts for why folk remedies were, and still are, thought to cure warts. The warts go away by themselves, but the folk cure gets the credit. Medical experts agree that the worst thing one can do is cut out the wart yourself. Besides endangering yourself by using nonmedical instruments, poor surgical technique, and unsterile conditions, you may infect yourself with the virus that caused the wart in the first place.

And that would be like throwing salt onto the wound.[13]

What are genital warts?

Genital warts are somewhat different than normal warts. All warts are caused by the human papilloma virus. But the human papilloma virus is a rather complicated animal. There are more than one hundred variations of this virus with more than thirty of these variations causing genital warts. According to Dr. June Reinish, head of the Alfred Kinsey Institute in Minneapolis, genital warts are smallish tumors outside and inside the genitals, but they do not

FEET ARE THE PART of the body most commonly bitten by insects.

always look like a wart. Indeed, some think genital warts look rather like cauliflower in appearance and color.

If untreated, genital warts can lead to abnormal cell problems in a woman's cervix, and this has been linked to cervical cancer in studies. The genital wart virus has also been linked to penile cancer.[14] Genital warts should be seen by the family doctor. There are ways to treat the condition, but treatment in women is difficult due to their location.[15]

Can kissing a frog or toad give me warts?

Of the many folklore remedies for getting rid of warts, one of the more memorable was from the MGM classic film *Huckleberry Finn* (1939), starring Mickey Rooney, which was of course based on Mark Twain's literary masterpiece *The Adventures of Huckleberry Finn* (1884): "You go to a cemetery where someone wicked has just been buried. You bring a dead cat in a gunny sack. At midnight, when the devil comes to take the soul of the wicked person, you swing the sack around your head three times and say, 'Devil take d' wicked, d' wicked take d' cat, d' cat take d' warts, I'm done wid ya'. And you throw the sack as far as you can."

Touching a frog or toad may not give you warts, but kissing or licking one *can* give you a "high." This belief is part of our folklore, in our nursery rhymes, and now (surprise, surprise) it's in our science.

Let's start at the beginning. The Frog Prince legend is part of the folklore tradition of most of the world. In fact, it is the first story in *Nursery and Household Tales* (1812) by the Brothers Grimm.

In one version of this thirteenth-century legend, a princess asks a frog to recover her lost ball from a lake in exchange for a trip to her castle. After the frog retrieves the ball, the princess forgets her promise, so the frog shows up unexpectedly in her bedroom. The princess keeps rejecting his advances but eventually kisses him, whereupon he turns into a handsome prince. Needless to say, they marry and live happily ever after.

In 1991, two U.S. doctors claimed they had discovered why this legend is so widespread. It all has to do with biochemistry.[16] Bufotenin, a chemical that produces hallucinations in humans, also has aphrodisiac effects, particularly in women. High amounts of bufotenin have been found in the skin of many common frogs and toads. Thus, according to Drs. David Siegel and Susan McDaniel of the School of Medicine and Dentistry at the University of Rochester, kissing or licking one of these warty creatures "can result in vivid hallucinations," romantic impulses, and sexual thoughts.

The two doctors add that this "biological property" was well known to people throughout the world for centuries. And it also explains why frogs and toads are often portrayed in folklore as "creatures of transformation or as intermediaries with other worlds."

Moreover, the two doctors note, "the magical moment" of the kiss "is, in fact, an aphorism known to many young women, that one must kiss many frogs to find one's prince."[17]

What are goosebumps?

Although many claim they get goosebump-type sensations from a wide range of stimuli, a goosebump is actually a muscular reaction to cold. When exposed to low temperature, the small muscles at the base of hairs start to contract. The effect is the formation of a mound around the hair. If the temperature stays low enough for long enough, a goosebump clearly forms and the hair stands erect. The medical name for goosebumps is *cutis anserina*. Goosebumps are reported from all around the world. It appears that no nation or ethnic group is immune.

Hair functions to protect the body against the harsh rays of the sun. This is probably why we have the most body hair on our head. The head needs the most protection, because the brain can be "cooked" inside the skull if unprotected on a hot, sunny day. This is particularly true if the body undergoes great exertion, as in "jogger's heat stroke." Hair also reduces friction problems in body movement. Perhaps that is why we have hair in our armpits and between our legs.

But hair also insulates the body against cold. In animals covered with hair, the rising strands of hair form a protective shield against cold. Cold air is trapped in and between hairs, instead of coming into direct contact with unprotected, sensitive skin.

Goosebumps may be a carryover from our earlier, more primitive primate existence. Although we humans have lost most of our body hair, compared with gorillas, chimpanzees, and orangutans, we still experience the same muscle contractions in reaction to cold as they do. But because we have less hair than our more primitive brothers and sisters, our goosebumps are more noticeable.

When an animal's hair stands on end, it looks ferocious. When ours stands on end, we just look cold.[18]

THE AVERAGE American uses fifty-seven sheets of toilet paper every day.

Why don't I get goosebumps on the palms of my hands or the soles of my feet?

This is an OBQ asked very often. It may also be the one requiring the shortest answer in this book: We get goosebumps only where we have hair.

Why does skin burn?

Our skin can burn simply because our bodies can be fuel, particularly our fat. The next time you roast some meat over an open flame, watch the drippings fall into the fire. If our fat was not fuel, there would be no reason for the body to store it.

What is meant by the terms first-, second-, and third-degree burns?

Contrary to common belief, first-degree burns are not the most serious. The degree of a burn refers to the extent to which heat has destroyed the layers of skin. A first-degree burn, such as a minor scald or sunburn, affects only the outer skin. These burns heal by themselves in a few days without scarring.

A second-degree burn is somewhat deeper and actually destroys some layers of skin, resulting in the formation of blisters. There is a moist, whitish surface color. If blisters are unbroken, they protect the injured area. Again, no scarring results and the skin regenerates after a few weeks. According to the Staff of the Royal Children's Hospital in Melbourne, a "deep second-degree" burn is somewhat more serious than a second-degree burn. In this case, there is a "moist white slough, red mottled" surface color.[19]

In a third-degree burn (also called a full-thickness burn), heat completely destroys the upper layers of skin, including accessory skin structures such as hair and sweat glands. The burn penetrates to the subcutaneous layer. The burned area takes on a dry, charred, white, leatherlike appearance, and normal skin does not regenerate. Such burns demand immediate medical attention.

Third-degree burns are often treated with skin grafts. Thin layers of skin are removed from an unburned part of the body and grafted onto the burned area. The donor site usually heals without problems, because only a thin layer of skin is taken. Skin grafts are not performed for merely cosmetic purposes. They may be necessary in third-degree burns because exposed subcutaneous tissue cannot heal fast enough to protect the body from infection and from loss of fluid.

Problems arise if the burn site is so extensive that there is not enough unburned skin on the body to perform the necessary grafts.

The skin used for the graft may be considerably smaller than the area of the burn to be covered. This is because a variety of techniques are available to enlarge the grafted skin. One technique is to cut the skin into filigree threads to cover the burn. Another technique is to mince the skin, place it in a nutrient solution, and grow skin sheets in a laboratory.

Pigskin and skin from cadavers can be employed as temporary grafting sources, but the body eventually rejects them.

Recent attempts to develop artificial skin have had mixed results. Among those developed in recent years is one consisting of a blend of substances, including shark cartilage. However, most doctors still believe that a graft of the patient's own skin is preferable because the body's immune system is more likely to reject any artificial skin as foreign.

Nevertheless, experts in this field believe that a rejection-free and fully tested artificial skin is likely to be available by the end of the century.

As research continues, it has in many respects only scratched the surface.[20, 21]

Why do men have nipples?

This is another frequently asked OBQ. In theory, we all could have fully functioning breasts capable of giving milk, but male breasts, including the nipples, do not get enough of the female hormone estrogen so they never achieve the ability to lactate (produce milk).

According to Dr. Bruce McEwen, a neuro-endocrinologist at Rockefeller University in New York and an international expert on sex hormones, "there are exaggerations in each sex based on early development and later hormonal influence. The nipple happens to be one of the features whose full development has been restricted in the male during early development so it never develops its full function, which is found in the female." He adds that the presence of nipples and other breast tissue in men illustrates the fact that the basic body plan of men and women is similar.

As far as we can tell, male breast tissue has no functional significance other than perhaps cushioning the heart and lungs from injury. However, nipples are considered erogenous in men as well as in women.

Nevertheless, Dr. McEwen points out that "the tissue is still there and can actually respond to female hormones in the male, as exemplified by the phenomenon of gynecomastia [abnormal enlargement of the breasts in men]." Gynecomastia occurs when there is an excess of estrogen. It is commonly seen in male alcoholics as well.[22, 23]

Could men ever give milk?

With the right hormonal treatment this is no problem at all. Dr. Jared Diamond, professor of physiology at UCLA School of Medicine, wonders why nature didn't give men the capability to suckle too. Wouldn't that double the number of people a baby could rely on for sustenance? In any case, Dr. Diamond writes: "Experience may tell you that producing milk and nursing youngsters is a job for the female mammal, not the male. But your experience is probably limited, and the potential of biology—and medical technology—is vast."[24]

Why do we blush?

This uniquely human physiological phenomenon has been fascinating and baffling since the first caveman committed the first social faux pas at the first human meeting place. Researchers say that by studying why we blush we can gain valuable insights into the complex and puzzling relationships between our mind, body, and society.

Blushing occurs when the small blood vessels that supply the skin widen, thus allowing an increase in blood flow. Blushers report a burning sensation in their face and often a full body tingle. In most cases, blushing passes within a few seconds to five minutes. According to Dr. Roger Dampney, a reader in physiology at the University of Sydney, blushing is a widespread phenomenon that science knows "surprisingly little about." Furthermore, "it is triggered by emotional stimuli," "higher levels of the brain are involved," and there is some anecdotal evidence that suggests blushing may not be confined to the face.

Blushing is one of only a few body changes triggered directly by the mind. It appears to be biologically driven rather than learned, but definitely socially induced. People do not blush in private. And you can make someone blush merely by accusing them of already blushing. Another curious thing about blushing is that people can blush and be made to feel embarrassed even when they have done nothing wrong. Simply being conspicuously different, good or bad, can raise a blush. Being complimented or overly praised is an example.

Even people who have been blind since birth are known to blush. And if you tell people they are starting to blush, chances are they will. In fact, researchers use this technique when attempting to induce blushing in order to study it. But at the same time, it is virtually impossible to make yourself blush. Chronic blushers sometimes

THE AVERAGE square inch of skin holds 650 sweat glands, twenty blood vessels, sixty thousand melanocytes (pigment cells), and more than one thousand nerve endings.

THE HUMAN body contains between fourteen and eighteen square feet of skin.

seek therapy to overcome the problem. They are often coached that when they feel a blush coming, they should try to make themselves turn as red as possible. It has been reported that in many cases this stops the blushing completely.

Despite theories going back to Darwin and Freud, no one is absolutely certain why only humans blush. Nevertheless, one simple explanation continually comes up: Humans are the only primates with facial skin completely exposed. Although other primates may also blush, blushing is visible only in humans. Another simple explanation is that blushing requires that a person have a sense of self, embarrassment capability, and the ability to judge themselves from the viewpoint of others. Because no other animal can do this to the degree that humans can, blushing is uniquely human.

Charles Darwin (1809–82) devoted an entire chapter to blushing in his book *The Expression of Emotions in Man and Animals* (1872). Darwin was the first to note that blushing is an exclusively human trait, universal among all peoples of the world and also characteristic of the blind. He argued that blushing was almost certainly an inherited trait and was triggered by the attention of others. Darwin wrote, "It is not the simple act of reflecting on our own appearance, but the thinking what others think of us which excites the blush."

In *Inhibitions, Symptoms, and Anxiety* (1926), Sigmund Freud (1856–1939) argued that blushing is a complex reaction that results from repressed sexual excitement and exhibitionistic wishes and that surfaces in the face of the fear of castration. It is an indirect way of conveying to others one's erotic urges. To Freud, a blush typified the internal subconscious power struggle between the id and the superego.

Even today, some Freudian psychologists still adhere to this basic point of view. For example, Dr. Carole Lieberman, a psychiatrist in Beverly Hills, California, puts it this way: "A gratified sexual wish is embarrassing. When a woman crosses a vent at an amusement park, her skirt flies up. She blushes. She may blush not simply because she is exposed but also because she had an inner wish to exhibit herself." Dr. Sydney Felman, a New York psychiatrist, puts it even more simply: "Men blush because they feel castrated, and women blush because they are not men."

Blushing is in many ways strangely contradictory. According to Dr. Murray Bilmes, a psychologist in Berkeley, California, "the striking thing about blushing is its implicitly mixed signals. A blush is a funny mixture of wanting to hide and at the same time wanting to attract someone." He notes, for example, that in a group of three when one person discloses to a second person that a third person has a secret passion, the third person blushes. Dr. Bilmes says, "Something has been exposed that they want to hide. Yet the reaction is partly to hide, but also to confirm it. There is a funny dialectic going on. That is a very important part of a blush." He adds: "Blushing is literally waving a red flag at a bull, a come on, a variation on the fight-or-flee response. Blushers want to hide, yet the blush draws attention to themselves."

Nevertheless, an additional theory for why we blush holds that blushing is actually an instinctive way to get back into the good graces of others. It is an attempt to avoid being ostracized from a group for breaching unwritten rules of society. The author of what is now known as the "appeasement theory" of blushing is Dr. Mark

Leary, a professor of social psychology at Wake Forest University. Dr. Leary first presented his theory at a 1990 meeting of the American Psychological Association.[25, 27]

We blush, Dr. Leary argues, "when we have done something that threatens our status in a group, when we have done something deviant. Blushing is an appeasement behavior to defuse a potentially ugly situation. Moreover, most animals have ways of doing this. For example, when other primates are threatened by a dominant animal they lower their eyes and get this really cheesy grin on their faces. Sometimes they avert their eyes, take on sheepish grins, or present their rear ends. It defuses the likelihood that they will lose status in the group or suffer aggression."

Dr. Leary notices that same "cheesy grin" on the faces of people who are blushing. While other primates use those instinctive gestures to allay aggression or avoid banishment, Dr. Leary claims that humans use them for the latter: "Blushing is saying to everyone else, 'Oops, I recognize that I've broken a social rule.' It is like a nonverbal apology—an instinctive acknowledgement that one has done something wrong—its purpose is to re-endear a person to a group in the face of impending exclusion."

Regardless of whether or not Dr. Leary is correct, there is some evidence that blushing works as Dr. Leary contends. At least blushing seems to create more empathetic feelings toward the blusher.

According to Dr. Rowland Miller, a psychologist at the Sam Houston State University in Huntsville, Texas, several studies suggest the power of the blush. For example, in one British study he cites, viewers watched a videotape of a clumsy shopper accidentally knocking over a supermarket toilet paper display. The researchers offered three alternative endings, asking the viewers what they thought about the culprit after each. In the first ending, the shopper merely flees. In the second, the seemingly unembarrassed shopper calmly rebuilds the display. In the third, the obviously mortified shopper blushes, looks around sheepishly, and eventually gathers up the wreckage. Dr. Miller notes that the views expressed the greatest degree of warmth toward the blushing shopper and concludes that a blush—the universal signal of human embarrassment—produces empathy while lessening hostility.[26]

Dr. Bilmes finds blushing very interesting: "Of all the possible things that separate animals from humans, the blush is our most distinctive characteristic. Why in the world did nature elaborate this thing?"[27-29]

HUMANS SHED and regrow outer skin cells about every twenty-seven days—almost one thousand new skins in a lifetime. By the age of seventy an average person will have shed 105 pounds of skin.

Why do I have fingerprints?

This OBQ is one of the more popular. "Vestigial" is the medical term given to a feature of the body that has no purpose. Although fingerprints have been accused by some of being vestigial, the reality is otherwise.

Fingerprints are visible parts of the rete ridges, where the skin's epidermis dips down into the dermis forming an interlocking structure. Our unique fingerprint and toe print configurations are due to the semi-randomness of ridge and dermal structure growth.

Fingerprints help us to grip and handle objects under a variety of conditions, on the same principle as automobile tires. A system

IF YOU ATE TOO many carrots, you would actually turn orange.

WHY CAN'T YOU tickle yourself? The brain's cerebellum warns the rest of the brain that you intend to tickle yourself. The brain is not fooled! Because your brain knows this, it ignores the resulting tickling sensations.

of troughs, ridges, and grooves has evolved to help channel water away from our fingers and toes. Although it doesn't seem like much, this results in a somewhat better gripping ability. Toe prints do the same thing. They help us keep our balance and prevent us from water-planing over smooth, wet surfaces.

Fingerprints also protect us against blisters. Fingerprints help alleviate the sideways stress that would otherwise separate the two layers of skin and allow fluid to accumulate in the space and thus form a blister.[30]

What turned Michael Jackson's skin white?

Dudley Moore had it. Steve Martin may too. Michael Jackson says he has it. Jackson claims that the treatment for this strange condition has bleached his skin white. This condition, called vitiligo, is surrounded by much mystery.

Vitiligo is a disorder in which the skin loses pigment when pigment cells called melanocytes have been destroyed. The result is that areas of the skin become white. The pigment loss occurs in patches, not over the entire body. The patches occur most often around body openings (such as the eyes), body folds (such as the armpits or groin), or exposed areas (such as the face and hands).

Vitiligo can strike either gender and at any age, but it usually shows itself before the age of twenty. Vitiligo is also quite common.

Perhaps one to two percent of the general population has it to some degree, although it may be confused with other skin problems. It is more common in people with thyroid conditions and certain other metabolic diseases. It is also more apparent in darker-skinned people. Yet most people who have vitiligo are in otherwise good health and suffer no symptoms other than the blotched areas of pigment loss.

Medical researchers are not sure what causes vitiligo. Some assert that the body may develop an allergy to its own pigment cells. Others argue that the cells may destroy themselves during the process of pigment production. Vitiligo is not contagious, nor is it in any way associated with leprosy. The old description of vitiligo as "white leprosy" has no basis in scientific fact.

Once vitiligo patches appear, there is no way to tell whether they will increase in size or number. In many cases, initial pigment loss will occur but then stabilize. In others, pigment loss can fluctuate. Psychological factors may play a part in such fluctuations, as many patients report that their initial or subsequent episodes followed periods of physical or emotional stress. A theory given serious consideration is that stress somehow triggers the depigmentation process in the human cell, among those who are already genetically predisposed. Sometimes, too, depigmentation patches may spontaneously repigment. Why this happens remains a mystery.

A common fear among vitiligo sufferers is that it is linked to skin cancer, that it is an early warning sign. However, there is no causal link between the depigmented patches and either cancerous or pre-cancerous conditions. But skin cancer patients sometimes develop vitiligo *after* their skin cancer symptoms appear. The reasons for this are unclear. Even more odd is that, in many skin cancer patients who develop vitiligo, the vitiligo seems to actually stop the spread of cancer. Why this happens is also rather baffling.

Vitiligo can strike anyone. In somewhat more than half of all cases there is a family history of the disease. Often too there is a family history of early graying of the hair as well. Statistically, early graying seems to foreshadow vitiligo and vice versa. Sometimes a vitiligo sufferer will not realize that there is a family history of the disease, and instead believe that there is merely a family history of early graying.

The good news is that vitiligo can be treated. In mild cases, makeup will cover the blotches without any treatment. Moderate cases respond to ultraviolet light, steroids, and especially to drugs like psoralen—alone or in combination. The goal of such therapy is to repigment white patches to a darker color. Repigmentation seems to work best when only a few small white patches exist.

IN ONE SQUARE inch of your hand you have nine feet of blood vessels, six hundred pain sensors, nine thousand nerve endings, thirty-six heat sensors, and seventy-six pressure sensors.

Presumably, in Jackson's case the repigmentation failed. When it fails in the most serious vitiligo cases, *de*pigmentation is attempted. Monobenzone is used to bleach the body's entire skin surface in order to give the patient at least an even, all-over color. Under a physician's careful instructions, monobenzone is applied two or three times daily until the bleaching is complete, and then twice weekly afterward.

Jackson is almost certainly using monobenzone because it is "the *only* treatment" to depigment, according to Dr. James Nordlund of the Department of Dermatology at the University of

Cincinnati School of Medicine. Dr. Nordlund adds that monobenzone is to be prescribed with care and only in "patients with very extensive vitiligo." Furthermore, the drug occasionally produces irritation, takes about six to twelve months to fully take effect, and has about a 75 percent success rate.[31]

More experimental treatments for vitiligo include the drugs clobetasol and tacrolimus. In a comparative study of these two drugs by Dr. Veronica Lepe and colleagues from the University of San Luis Potosi in Mexico, it was found that both drugs worked about as well in treating the condition but that tacrolimus was preferred for sensitive skin areas.[32]

If Jackson stops his monobenzone treatments, his previous color will return. In fact, Michael Jackson could turn back to black whenever he wants.[33]

Why is a baby's skin so soft?

Recent research points to vernix as an additional factor in making baby skin soft. Vernix is a white, cheesy substance that coats infants during the last few weeks before birth. In modern hospitals, whatever vernix is still on the baby at birth is wiped or washed off.

However, a study for the Skin Sciences Institute of the Cincinnati Children's Hospital found that "newborn skin with vernix left intact is more hydrated, less scaly, and undergoes a more rapid decrease in pH than with vernix removed." Therefore, according to Dr. Marty Visscher, executive director of the institute, "these beneficial effects of vernix suggest that it should be left intact at birth."

Vernix is composed of a complicated mix of lipids, proteins, and water that is nature's perfect skin enhancer. It's possible that future skin rejuvenation products will include the chemical constituents of vernix. Nature so often does it so much better.[34]

chapter 7
the **hair** and **nails**

I n "The Rape of the Lock," the poet Alexander Pope (1688–1744) wrote the immortal lines "Fair tresses man's imperial race ensnare, And beauty draws us with a single hair." If ol' Alex didn't have most of us in mind with that one, he was awfully spot-on just the same. Hair has been an important part of our human life and culture—the Samson story from the Bible, the Rapunzel fairy tale, and on and on.

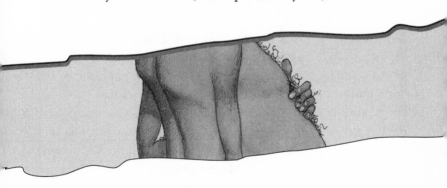

Why does my hair turn gray?

This is a frequently asked OBQ. In the vast majority of cases, age and the erosion of melanocyte-functioning cause hair graying.

The graying process starts deep in the outer layer of the skin (epidermis) all the way to the inner layer of the skin (dermis). Each of the one hundred thousand or so hairs on the average human head is controlled by a hair bulb located below the hair follicle. A variety of complex chemical substances that are composed mainly of keratin and that create each hair are channeled through the hair bulbs.

Millions of melanocytes (protein-producing pigment cells) located at the hair roots and in the epidermis produce chemicals that determine the coloring of both hair and skin.

The melanocytes yield the color by treating the hair at the follicle, and eventually the hair grows long enough to reveal the color. Once the hair has been pigmented by the melanocyte's action, the color cannot be altered. This is because the pigmentation does not just coat the keratin body of the hair but is infused.

Those who suffer from albinism usually have a normal number of melanocytes but, due to a genetically caused deficiency, lack the chemical means to trigger pigment production. In some people, only a small area of skin lacks functionless melanocytes. This produces white spots or streaks in an otherwise dark head of hair.

THE MIDDLE finger boasts the fastest growing nail.

HOO SATEOW of Chang Mai, Thailand, has the longest head of hair in the world. His locks measure sixteen feet, eleven inches. His brother Yee's head of hair is sixteen feet long.

The pigmentation chemical, melanin, has two basic components that predispose a hair to be dark, light, or some hue in between, depending on the proportion of each pigment. Graying occurs when there is an age-related shutdown of the melanocytes, which usually occurs gradually over several years. Although graying can occur quickly, you cannot turn gray overnight. Your hair does not fall out that fast. Although some people report that they turned gray overnight, closer examination invariably shows that they were turning gray slowly but just didn't notice the change. Their awareness of being gray is what happened overnight.

About one hundred hairs a day are lost from natural attrition. With age, the older dark hairs fall out leaving a greater proportion of newly created white hairs. As white hairs gain the majority, the grayness appears to increase. Thus, grayness is an optical illusion created by the mixture of the remaining dark hairs and the newer white hairs.

Graying is almost certainly largely genetically determined. Nevertheless, stress and worry can probably influence the rate at which a person turns gray.

Men and women gray in slightly different patterns. Women gray slightly faster than men. Premature graying affects about 25 percent of all people by the time they are twenty-five years old. Production

MEN WITH mustaches may be allergic to their own lip hair. That's because mustaches can harbor airborne pollens that trigger allergies.

IN CERTAIN PARTS of Romania, women with mustaches are the most sought after for wives because their hairy upper lip is thought to be a symbol of fidelity and fertility.

of the first gray hairs begins at about age fifteen. Ironically, the colorizing cells often speed up pigment production as we age, so hair sometimes darkens temporarily just before the pigment cells die.[1]

Thyroid disorders are a common cause of premature failure of melanocyte functioning.[2] Diseases affecting the pituitary gland probably reduce hair colorization, as can interruptions of hormone production in the testicles or ovaries. Diabetes can upset melanocyte functioning, as can severe malnutrition. Premature graying has also been associated with a possible increased risk of heart disease. And evidence suggests that in cases of pernicious anemia a deficiency of vitamin B_{12} may influence melanocyte functioning and thereby influence graying.[3]

Why do some people have curly hair while other people have straight hair?

The pattern of hair can be naturally straight, curly, or somewhere in between. Genes ordain not only hair color but also hair pattern. Small communities that are endogamous (marry inside the group) will, over many generations, evolve a single hair pattern. An example of this is the San people (sometimes called the Kung Bushmen) of the Kalahari Desert of Southern Africa. All San have "peppercorn" hair. As this vivid term implies, their hair is very, very tightly curled.

How does a permanent make my hair curly?

The chemicals in a permanent make hair curly by first breaking down and then re-forming the actual chemical bonds of hair.

Hair is composed mostly of a protein called keratin, which contains large amounts of sulfur. Chemical bridges connect sulfur atoms from one protein molecule to nearby sulfur atoms in neighboring molecules. Like rivets holding the steel girders of a skyscraper together, these chemical bridges hold the keratin molecules together in a structure that gives each hair its characteristic shape.

One can bring about a temporary alteration of the hair's natural pattern in several ways and for varying lengths of time. For example, by tightly wrapping straight hair around curling irons, one can hold the curl for a short time. However, the alignment of the hair's proteins will eventually force it back into its naturally straight shape pattern.

When you get a permanent, the proteins in the hair are allowed to realign so that they favor a curly shape instead of the straight shape. The hairdresser first winds the hair around the curlers and then applies a chemical that breaks down the bridges between the hair proteins. Just as steel girders can slide when their rivets are removed, the proteins can now slide with respect to one another. Freed from their former restraints, the proteins settle into positions that conform to whatever shape the hairdresser determines. If the hair is wrapped around curlers, curly hair the result.

IN CHINA, wealthy men used to grow their fingernails to astonishing lengths, so long that their nails were covered by a silk bag to protect them. In this way a wealthy man demonstrated his status, showing that he had servants to do everything for him.

BY THE AGE OF twenty-one, one in every twenty young males will find that their hairline is receding.

151

Nevertheless, there is one final step. The reordering must be "locked in." The hairdresser eventually washes away all the bond-breaking chemicals and adds a different chemical solution that causes new bridges to form. This fixes the proteins in their new, curly configuration. Even wind, sun, and shampooing cannot alter this rearrangement. Only after the hair grows out again does it resume its original genetically determined pattern.

As powerful a solution as a permanent is, it cannot alter our genes.[4]

Why do only men have beards?

This is a matter of hormones. Facial hair is linked to testosterone, the male hormone. Genes determine our level of testosterone. Males have more testosterone than females, and hence usually more facial hair. Studies have shown that bearded men are "rated significantly higher on masculinity, aggressiveness, dominance, and strength" by both males and females.[5]

The ancient Iraqis were the first true hairstylists. Then known as the Assyrians, and inhabiting what is now northern Iraq, the skill of Assyrians at cutting, curling, layering, and dyeing hair was legendary throughout the Middle East. Vidal Sassoon would have had every right to be green with envy, except that this was some 3,500 years ago. Nevertheless, this ancient Assyrian craft grew out of a love for hair—an obsession with hair—especially for soldiers in battle.

The Assyrians cut hair in graduated tiers, so that the head of a fashionable general was as neatly geometric as an Egyptian pyramid—and somewhat similar in shape, although upside down. Longer hair was elaborately arranged in cascading curls and ringlets that tumbled over the shoulders and onto the chest.

Hair was oiled, perfumed, and tinted. Men cultivated a neatly clipped beard, beginning at the jaw and layered in ruffles down far enough to completely cover the neck. Kings, generals, and lieutenants had their abundant, flowing hair curled by slaves using a fire-heated iron bar—the first hair curling iron known to history. No self-respecting warrior would dare enter battle without ensuring that his hair was properly done. Hair, like armor, had to be worn properly and with distinction. If a warrior fell, he fell with style.

Indeed, the Assyrians were so obsessed with hair that they developed hair-styling to the exclusion of nearly every other cosmetic art. Law even dictated certain types of coiffures according to a person's rank in military and civilian life.

One rule prevailed: Baldness, full or partial, was a sign of weakness and impotence—the mark of the eunuch. To the Assyrians, baldness was an unsightly defect to be concealed in public by wigs.

Hairstyling was extremely important to Assyrian women too. As in ancient Egypt, high-ranking Assyrian women appearing at the royal court would appear only after donning stylized fake beards, to assert that they could be as authoritative as men. The beard was the symbol.

Only later, with the ancient Greeks and Romans, did beards and long hair fall out of favor for the battlefield. Alexander the Great

SOME PEOPLE have no hair at all. This condition is called *alopecia universalia*.

FOR REASONS not fully understood, hair grows faster in the morning than at any other time of day.

THE AVERAGE human hair can support nearly four and a half pounds of weight.

demanded that his soldiers cut their hair and beards before battle, and the Romans shaved before battle too, with crude razors. For hand-to-hand combat long beards and long hair were seen as undesirable because they could be used to pull a man down.

In fact, the Romans used the word for "beard" (*barba*) as the term for their enemies in battle. Hence, we have today the word "barbarian" as well as "barber."

Why do some people pull their hair out?

Many of us pull or twirl our hair, sometimes merely as a habit developed in childhood. But some people compulsively pull their hair out, often including their eyebrows, eyelashes, underarm hair, and even pubic hair. This bizarre condition is called trichotillomania.

Mental health professionals usually classify trichotillomania as an "impulse control disorder." Although chronic hair-pulling will not kill anyone, some individuals cause themselves injuries and pain and risk disfigurement if they persist in giving in to this impulse. Moreover, the condition is associated with high rates of what is termed psychiatric "co-morbidity"—that is, other psychiatric conditions occur along with hair-pulling, although no causal connection can be proven. For example, trichotillomania is associated with marked emotional distress and dimished self-esteem. But rather than one causing the other,

IF AN AVERAGE man never shaved, his beard would grow to about ten yards.

THE AVERAGE MAN has about thirty thousand whiskers on his face and will spend about 3,350 hours in his lifetime shaving them.

they all may be reinforcing one another. People who frequently pull their hair do not necessarily have trichotillomania. The internationally accepted standard of the *Diagnostic and Statistical Manual of Mental Disorders* (DSM-III-R), published by the American Psychiatric Association, recommends that doctors should look for five signs or clinical criteria before diagnosing trichotillomania:

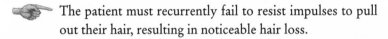 The patient must recurrently fail to resist impulses to pull out their hair, resulting in noticeable hair loss.

The patient must experience an increasing sense of tension immediately before pulling out the hair.

The patient must feel gratification or a sense of relief when pulling out the hair or immediately afterward.

The patient must have no preexisting inflammation of the skin.

The patient's hair-pulling must not be a response to any delusion or hallucination.

Nevertheless, although most of us probably know a hair-puller (often a child), trichotillomania has for the last thirty years generally been considered a rare condition. For example, Dr. S. Muller of the Mayo Clinic in Minneapolis reported that there are only fifteen cases a year among the thousands of patients seen for various complaints at the world famous clinic. Moreover, it is widely believed that trichotillomania is more common in women than in men—at a ratio as high as nine to one. Indeed, Dr. Muller found that three out of four trichotillomania sufferers are women.[6]

But in 1991, three psychiatrists from the University of Minnesota claimed that conventional wisdom about trichotillomania was wrong. Drs. Gary Christenson, Richard Pyle, and James Mitchell found that "trichotillomania may not be as rare as previously suspected and may affect males almost as often as females."[7] They based this conclusion on their study of 2,579 university students that showed that 1.5 percent of men and 3.4 percent of women had visible hair loss as a result of trichotillomania even though less than half of these chronic hair-pullers met all five trichotillomania criteria. When all the criteria were met, however, just as many men as women qualified as true trichotillomania sufferers. Furthermore, the researchers stress that theirs was not a clinical population of patients but a more general group of people, which is all the more surprising.

If these doctors are right, then trichotillomania probably occurs more often in the general population than among people treated for mental illnesses. And even though men have shorter hair—the common reason given for exhibiting less trichotillomania—they may suffer from it nearly as often as women.

On my weekly radio segment on 3AW in Melbourne, the head of one of three "tricho" support groups in Melbourne (who prefer to remain unidentified) claimed that perhaps one in twenty Australian women and one in thirty Australian men suffer from this disorder. This was based on anecdotal evidence and discussions with clinicians.

Why do people engage in chronic hair-pulling? We don't know. It is one of the mysteries of human behavior, if not of the body.

Dr. Judith Rapoport of the National Institute of Mental Health in Washington, D.C., suggests some possibilities. For example, trichotillomania may be the chosen form for a compulsion because it has important symbolic value. In classic psycho-analytic literature, hair can symbolize a number of things, such as beauty, femininity, virility, and physical prowess.

In addition, Dr. Rapoport writes, "It is also a bisexual symbol and sexual conflicts are said to be displaced to the hair." Haircutting, pulling, or plucking is believed to symbolize castration or the rendering of oneself as infertile or impotent. For the victims of the Nazi concentration camps, for example, the shaving of the head was considered to be one of the most excruciating of tortures and debasements.

Moreover, Dr. Rapoport relates that a psychoanalytic essay by Dr. Edith Buxbaum on the fairy tale "Rapunzel" suggested that the cutting of Rapunzel's hair represents separation from and loss of the mother. Furthermore, in various Hindu cultures shaving the scalp is associated with mourning, and in some areas of India, hair is plucked before a person enters a life of penance. The Christian monastic life too has a long tradition of shaving the head.

Although so far we have only poor explanations for what causes trichotillomania, we do know that those who suffer from it are utterly miserable. Fortunately, the news is far better about the *treatment* of trichotillomania. Talk therapy often achieves marked improvement,

DURING Abraham Lincoln's U.S. presidential campaign in 1860, a Democrat named Valentine Tapley from Pike County, Missouri, swore that he would never shave again if Lincoln was elected. Tapley kept his word, and his chin whiskers went unshaved from November 1860 until he died in 1910, attaining a length of twelve feet, six inches.

and support groups are very helpful. Certain drugs, such as clomipramine, often have dramatic results. In one study, nine out of ten trichotillomania sufferers got "much better" through the use of this potent antidepressant.[8]

Drug therapies are continually being improved, and there is hope for people with trichotillomania. They need no longer tear out their hair—in more ways than one.[9]

If I lose my hair, will it ever grow back?

While a few of us pull our hair out, many of us want to put it back. Losing hair as a result of age is not just a male problem. About 30 percent of women also suffer from age-related hair loss by the time they reach the age of forty. This is called "pattern alopecia."

A new experimental product may help us retain our youthful hair long after it would normally lose its body and sheen. Actually, the new product is already a popular prescription drug that is, oddly enough, a diuretic. It is called spironolactone.

Research at the Philip Kingsley Trichological Centre in London claims that spironolactone "greatly improves" pattern alopecia. A Kingsley endocrinologist gave oral doses of spironolactone to six women between the ages of thirty and forty-five who were suffering from moderate to severe hair loss. In four of the women the drug stopped further loss, while in two others it unexpectedly promoted hair growth. The only side effect of spironolactone seems to be a slight menstrual irregularity.

Spironolactone is known to reduce the activity of testosterone and other male-associated hormones, and female pattern alopecia is triggered by a sensitivity to these hormones.

The researchers expanded the trial to include twenty-five more women and twelve balding men. They are now testing spironolactone on human subjects. The initial results continue to be promising.

Dr. David Kingsley, founder and director of the Trichological Centre, claims: "We haven't found the magic oil that will grow full heads of hair, but clearly spironolactone is well worth investigating further."[10]

Why do I bite my nails?

About 25 percent of both Americans and Australians bite their nails. Most bite just their fingernails, but some bite their toenails as well. (The medical term for nails is *unguis,* and the medical term for nail-biting is *onychophagia.*)

It is unclear why so many people do bite their nails. The most common theory is that nail-biting is stress related. This notion holds that nail-biters alleviate built-up, everyday tension through the unconscious and unintentional biting or peeling of their nails. Thus, nail-biting functions similarly to scratching the head, pulling the hair, cracking the knuckles, and so on. Another theory, a neo-Freudian theory heard less often these days, is that nail-biting is a substitute for masturbation, that nail-biting emerges because it is less heavily repressed and less socially taboo than masturbation. Still

DURING THE time of Peter the Great of Russia, any Russian man who had a beard was required to pay a special tax.

FOR A WHILE, composer and pianist Frederic Chopin (1810–49) wore a beard on only one side of his face. He explained that because he sat sideways the audience saw only one side of him.

another view holds that nail-biting reflects nutritional deficiencies, especially in cases where the nail shaving is swallowed. Nails (and hair) are made of a particularly strong protein called keratin.

Nail-biting can be unsightly, but it can also have health consequences. According to Drs. Peter Samman and David Fenton, nail-biting can make a person vulnerable to skin infections. In cases of severe nail-biting, when the nails are bitten all the way down to the whitish, semicircular "half moon" region near the cuticle, permanent nail deformity can occur.[11]

Nail-biters normally try numerous remedies to curb the biting or peeling urge: bandages, aversion therapy using hot sauce or bitter-tasting nail coatings, relaxation tapes, hypnosis, chewing gum, and many others. "Sadly, these remedies seldom work," says Dr. Richard Scher, professor of dermatology and director of the Nail Research Center at Columbia University in New York. Dr. Scher, along with Dr. C. Ralph Daniel, is the author of perhaps the most influential work on this topic.[12]

THERE ARE typically one hundred thousand hairs on the adult human head.

BLONDES HAVE more hair than dark-haired people.

Dr. Scher notes that because nail-biters usually chew their nails unconsciously they cannot stop until they develop a keen awareness of when and under what circumstances they are most prone to bite. Indeed, "they should cultivate an awareness of where their hands are at all times."

It is self-defeating for parents to punish children who bite their nails, because punishment tends to increase tension and make matters worse—perhaps resulting in even more nail-biting. Instead, parents should point out to children when they are biting their nails and in a supportive manner encourage them to be more aware of what their hands are doing. Reducing stress in the child's life also helps.

Dr. Scher wants to dispel several myths exist about nails and nail-biting:

☞ It is not true that manicures make no difference. "To the contrary, nail polish makes nails more attractive, and the better looking our nails, the less likely we are to abuse them."

☞ It is not true that nail polish is for women only. "In fact, a growing number of men are now having their nails manicured. And clear polish is virtually undetectable."

☞ It is not true that gloves make no difference. "Covered nails are not only less tempting than uncovered nails, but also are harder to reach if an urge to bite proves irresistible. . . . For maximum protection, wear gloves at all times—day and night—until the nail-biting urge dissipates. If you have the urge and are not wearing gloves, lightly clench your fists until it passes."

☞ It is not true that psychotherapy does not work. "In particularly severe cases of nail-biting, short-term therapy may be helpful. In some cases, tranquilizers may be prescribed."

SIDEBURNS, the whiskers adorning the side of a man's face, got their name from an American Civil War general who wore an exaggerated version of the style, shaving only his chin. However, the man's name was not General Sideburn, but General Ambrose Burnside.

IF A MAN HAS A long vibrissae he has a long mustache.

INTELLIGENT people have more zinc and copper in their hair.

THE NAIL ON THE thumb grows slowest.

A FINGERNAIL OR toenail takes about six months to grow from base to tip.

The practice of polishing, painting, or otherwise adorning nails goes back to the ancient Chinese some five thousand years ago. In ancient China, both women and men used a paint made of beeswax, egg whites, gelatin, and gum arabic. At roughly the same time, the ancient Egyptians adorned their nails too. Egyptian women and men of high social standing used henna to stain their nails a rich, red-orange—the deeper the red, the more exalted and important the individual.

Like hair, nails are composed of dead tissue. That is why it does not hurt when hair or nails are cut—only if pulled. The living growth is deep within the skin.

Fingernails grow about one and a half inches a year. Fingernails grow two to three times faster than toenails. Few factors can speed the rate of nail growth, but we do know that nails grow faster during the day than at night, and grow thicker as we age. Warmer weather speeds the rate of growth slightly because heat increases the rate of all metabolic processes.

Generally, nails of males grow slightly faster than nails of females, but women can expect a slight spurt in their normal nail growth rate just before menstruation and throughout pregnancy—a response to hormone activity.

Nails do not grow after death. It merely looks that way, because the skin dries, shrinks, and retracts from the ends of fingers and toes.

Oh, and yes, death is a proven cure for nail-biting—although somewhat extreme.[13]

Why are some people so hairy?

Excessive hair occurs in 10 percent of adolescent girls and an equal number of women during the childbearing years. In such cases, the hair grows on the face, chest, back, buttocks, linea alba (the line between the navel and the pubic region), and elsewhere, violating current fashion standards. It is a little easier for males since current fashion allows males to be a little hairer—although not much. In any case, both genders can panic over too much hair.

Hirsutism is a medical condition in which there is an excessive amount of body hair. Of course, just what constitutes excessive is highly subjective. There is no standardized medical definition. Nevertheless, about one in ten girls and young women have "a significant degree of hirsutism," according to Dr. Gordon Senator, an endocrinologist at the Royal Hobart Hospital, writing in the *Australian Dr. Weekly.*[14]

Dr. Senator draws on one of the few studies of hirsutism rates. In a 1964 United Kingdom report in *The Lancet* by Dr. E. McKnight of the University of Wales, a survey of four hundred women students found that 26 percent had "terminal hair" on the face. In these individuals, more hair appeared on the upper lip than on the chin. Furthermore, facial hair was clearly noticeable to

FINGERNAILS grow nearly four times faster than toenails.

IF A HUMAN HAIR were as thick as a nylon rope, it could support a train engine.

an impartial observer in 10 percent of the women. Finally, in 4 percent of the women, hair was sufficient enough to require treatment.

Virilization is different from hirsutism. Virilization is the induction of increased body hair but also deepening of the voice, increase in muscle mass, development of a hairline typical of the male forehead, stimulation of secretions and proliferation of the sebaceous glands often leading to acne, hypertrophy of the clitoris (the clitoris becomes enlarged), oligomenorrhea (menstrual periods become erratic and flow is reduced), and/or amenorrhea (periods cease).

Both hirsutism and virilization are caused by an overabundance of testosterone, the male hormone. Usually this is genetically based, but sometimes diseases or other medical conditions may be involved. Polycystic ovary disease is perhaps the most common of these.

In the classic text, *Hirsutism,* by Drs. P. Mauvais-Jarvis, F. Kuttenn, and I. Mowszowicz, three endocrinologists point out that there is "a direct relationship between the extent of hirsutism and the level of plasma testosterone in hirsute women." In general, the more male hormone, the greater the degree of virilization. However, the three doctors observe that "in most cases of hirsutism, plasma testosterone is either only slightly increased or within the normal female range."[15]

Hirsutism does not indicate a lack of femininity. Femininity, based on the current standard of

THE LENGTH OF the finger dictates how fast the fingernail grows, so the nail on your middle finger grows the fastest.

THE HAIR AND nails are the only two places in the human body without red blood cells. All other organs have blood flowing through them constantly to provide the necessary oxygen and nutrients.

feminine beauty dictated by the fashion industry, is highly subjective. What is considered beautiful in one historical period may not be in another. In fact, during Victorian times at least a moderate degree of hirsutism in a lady was considered quite fashionable and very attractive. Men's views about women's body hair are quite diverse, as are women's varying preferences for facial or chest hair on men.

The degree of hair growth varies among ethnic groups. Northern European women tend to have less hair than Southern Europeans. Asian women have less still. Moreover, birth-control pills can cause extra hair growth by altering normal hormone levels.

Should hirsutism be severe, it may be treated. According to Drs. Stephen Judd of the Flinders Medical Centre in Adelaide, and John Carter of the Concord Hospital in Sydney, writing in the *Medical Journal of Australia,* cosmetic treatments include "bleaching, plucking, depilatory creams, waxing, shaving, and electrolysis."[16] Various hormone therapies can also be used. Hormonal treatment of hirsutism should not affect libido or other aspects of sexuality.[17]

Hirsutism does not become worse if a woman removes the hair by shaving, because hair is created far below the skin's surface. Shaving does not affect the rate of hair growth or its coarseness.

If there is any doubt, concern, or anxiety about hair growth, see your family physician to allay all fears.

At times a young woman may turn *to* a man, but never *into* a man.[18]

Do humans get hair balls like cats?

N ot exactly, but your own hair can kill you. Hair bezoars (trichobezoars) are pathological masses of swallowed hair that form, harden, and become lodged in the human gastrointestinal tract. We have the capacity to form bezoars, as do such animals as goats and Persian cats. Human bezoars can cause abdominal obstruction, hemorrhage, and perforations. Death results in 30 percent of diagnosed cases left untreated. A bezoar-related death is often prolonged and excruciatingly painful.

Hair bezoars occur occasionally in people who chew their hair, most frequently young girls and women. People who chew their hair inevitably swallow some too. If this is done habitually, a hair bezoar can result. The person suspects nothing until symptoms emerge. Hair bezoars are less of a health problem these days because hairstyles for both males and females are shorter.

Medical treatment for human bezoars can include endoscopy (viewing the bezoar through an inserted small, flexible fiberoptic tube), injecting the bezoar with an enzyme solution to fragment it, or surgery.

There are two other types of human bezoars: phytobezoars and trichophytobezoars. A phytobezoar is composed of vegetable matter, such as citrus-fruit pulp, fibers, skins, and seeds, and occurs when too much of such material is swallowed. Of all fruits, the skin of the persimmon seems to be the most prominent in forming phytobezoars. A trichophytobezoar is a combination of hair and vegetable matter, hence its name.

Articles on bezoars surface in the medical literature from time to time.[19] In one of these, three Belgian doctors used the colorful name "Rapunzel Syndrome" to describe an unusual case of the trichobezoar in a fourteen-year-old girl who loved to eat her stuffed toy animals as well as the family's carpet.[20]

the **hair** and **nails**

If it seems bizarre that humans would swallow such things, remember that there are more than thirty cases in the medical literature of humans swallowing just one everyday household object—the toothbrush.[21] Perhaps one of the weirdest things swallowed was a condom filled with beer. This unfortunate incident occurred at the Munich Octoberfest.[22]

Four doctors from the Department of General Surgery at the Faculty of Medicine of Gaziantep University in Turkey reported on their successful attempt to remove a bezoar the size of a basketball from the abdomen of a twenty-four-year-old woman.[23]

A MAN'S HAIR grows faster when he believes he's about to have sex. This is less so for women.

MEN LOSE about forty hairs a day. Women lose about seventy hairs a day.

A landmark article on human bezoars was written by Dr. Randolph Williams, visiting surgeon at the Royal Adelaide Hospital.[24] Dr. Williams writes that bezoars (from the Arabic *badzehr*, meaning "poison antidote") were once highly prized throughout Europe. They were thought to contain great medicinal properties. Indeed, the most valued bezoar of all was that obtained from the fourth stomach of the Syrian goat. Down-market, second-rate bezoars from other animals and humans were often substituted as counterfeits, because Syrian goat bezoars were always in short supply. During the Middle Ages, trade in bogus bezoars flourished, so many tests were devised to detect the fakes.

Bezoar stones were carried as charms to ward off illness and poisoning. For example, at the king's table, guests would be served wine in a jeweled goblet. Attached to each goblet, hanging from a small chain so it would be handy for dipping (similar to a teabag of today), would be an equally bejeweled bezoar. In addition, bezoars were often pulverized to a fine powder and administered internally as a treatment for many medieval maladies.

A bezoar stone was once set in gold and included in an inventory of crown jewels belonging to Queen Elizabeth I. The Queen's high esteem for bezoars, Dr. Williams says, "attested to its popularity with English royalty."

Bezoars remained in medical use until the mid-eighteenth century, when they were finally superseded by more effective treatments. Despite the high cost of medical care these days, and the movement toward alternative and traditional folk remedies, it is unlikely that bezoars will make a comeback as a standard form of therapy.

But if you notice a Syrian goat grazing in back of your doctor's office . . .[25, 26]

If hair is not alive, why does my physical or emotional state change the condition of my hair?

Even though hair is not alive, how it looks depends on secretions from the sebaceous glands at the base of the hair follicles. Too little of the secretion, and the hair becomes dry and brittle—damaged. Too much of the secretions, and it becomes greasy and lifeless. (Sounds like a shampoo commercial, doesn't it?) Sebaceous secretions are affected by age, general health, hormone flow, and even one's emotional state. Pregnant women often notice changes in their hair. These changes are temporary, perfectly normal, and due to the natural hormonal changes occurring in pregnant women's bodies.[27]

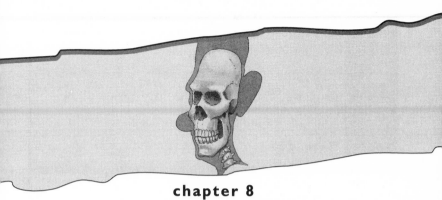

chapter 8

the **skeleton, bones,** and **teeth**

T he human skeleton is a wonder of engineering. You could compare it to a strong building. For example, pound for pound, your femur (thighbone) is stronger than reinforced concrete. When you walk fast, your femur resists an average of 400 to 450 pounds per square inch. Most of the world lives in houses less sturdy than that!

Bones consist of calcium and phosphorus so densely packed that the crystalline pattern of bone resembles that of diamonds, the hardest natural substance known. And let's not forget the strength of teeth. Tooth enamel is the hardest substance produced by the human body. It is so hard that a dentist's drill must turn one million revolutions a minute just to cut through it.

How many bones are in the human body?

You're bound to be asked this sometime, and you're right if you answer 208. The answer is not likely to change in the next few years, so you go ahead and memorize it.

Where are the various bones located? Well, more than half of your bones are in your arms and legs, if you count the hands and feet with them.

The bone breakdown goes like this: Sixty in the upper extremities (arms and hands), sixty in the lower extremities (legs and feet), twenty-six in the vertebrae (backbone), twenty-four vertebral ribs, twenty-two in the skull, six in the ears, four in the pectoral girdle (collar and shoulders), three in the sternum (breastbone), two in the pelvic girdle (hips), and one in the throat.

You're also bound to be asked how many teeth are in the human body. Go to the head of the class, if you correctly answer thirty-two.

How heavy can a person be?

Although scientists debate this, there is probably no limitation on how large humans can become. But already we know humans are capable of doing things in big ways—very big ways.

The world's heaviest human ever was Jon Brower Minnoch (1941–83), who weighed more than 1,383 pounds at his heaviest. In

March 1978, the six-foot-one-inch Minnoch checked into the University Hospital in Seattle, saturated with fluid and suffering from heart and respiratory failure. Twelve firemen were required to move him, and two hospital beds lashed together were required to hold him. Dr. Robert Schwartz, a consultant endocrinologist, calculated Minnoch's weight based on his intake and elimination rates.

Minnoch recovered and reduced his weight to a low of 473 pounds, but much of the weight returned. When he died he weighed 796 pounds.

The world's heaviest surviving human baby by vaginal delivery was a boy weighing 22.4 pounds, born in 1954 to Signora Fedele of Aversa, Italy. The heaviest baby by cesarean section delivery was also a boy, who weighed the same, born in 1982 to Christina Samane of Sipetu in the Transkei, South Africa.

THE THUMB IS considered to be a finger.

IF YOU CAN'T stop moving your fingers and toes, you suffer from athetosis.

SCAPULAMANCY is the method of fortune-telling involving the study of cracked shoulder bones.

Larger babies have been stillborn. The largest weighed 29.1 pounds and was born in Effingham, Illinois, in 1939.

What is surprising, too, is that a baby is not the heaviest thing a human body can produce. A cyst can be far heavier. On 24 October 1991 a gigantic 302-pound cyst was removed from the ovaries of a thirty-six-year-old California woman. When the massive abdominal growth was removed, it had to be carried out on its own stretcher. The woman is now fully recovered and leading a normal life.

Dr. Katherine O'Hanlan, a specialist in gynecological oncology, led the surgical team at the Stanford Medical Center in Palo Alto.

Dr. O'Hanlan said that the ovarian growth was by far the largest cyst in history. In fact, it outweighed the next largest by more than 109 pounds.

The patient was five feet nine inches tall and weighed 209 pounds immediately following the operation. Because of the great weight of the noncancerous cyst, she had been bedridden for two years before the surgery. The cyst, which had grown to more than one yard in diameter over a ten-year period, was quite unusual, because most ovarian cysts grow to no more than a half an inch or so. Dr. O'Hanlan says, "The patient had read about the difficulties of removing cysts much smaller than hers, and was wisely hesitant to have surgery. She was afraid of losing her life. But things had just gotten to be unmanageable." The delicate operation took more than six hours. According to Dr. O'Hanlan, the mass protruding from the woman's lower abdomen was made up of many small cystic structures fused together "like a water balloon that contains many other balloons." It "had to be carefully trimmed away from the abdominal wall and the bowels. We rolled it onto a stretcher so it could be sectioned or cut up to be examined under the microscope. None of us could lift it."

In May 1993, Mike Goodkind, a spokesperson for Stanford Medical Center, said that the patient "was very happy" and has a "skinny abdomen now."

THE WORLD'S shortest married couple was not "General and Mrs. Tom Thumb," Chares and Lavinia Stratton of P.T. Barnum fame. The shortest couple are Douglas Maistre Brager de Silva and Claudia Pereira Rocha, Brazilians who were married on 26 October 1998 and measure thirty-five inches and thirty-six inches respectively.

Previous to this, the largest cyst ever removed from a patient was a 193-pound growth removed in Maryland earlier the same year.

The biggest foreign creature that can live inside the human body is a parasitic worm. There are some eight hundred thousand species of such nematodes, the largest of which can grow to several yards in length but usually inhabits only the placenta of the sperm whale. Theoretically, if a human ate the whale, the worm could infect the human—at least for a while.

Of course, bacteria and viruses live inside us too. The largest bacterium known to exist (*Epulopiscium fishelsoni*) is a recently discovered, single-cell organism that normally lives in the bowels of a fish (*Acanthurus nigrofuscus*) inhabiting the waters at Lizard Island, Queensland, Australia.[1] Until a few years ago, bacteria were believed to be so small that none could be seen with the naked eye. However, this new mammoth bacterium is about the size of a hyphen in a newspaper. Dr. Ester Angert, a researcher at the University of Indiana in Bloomington and co-discoverer of the giant bacterium, said: "It's so huge that we could stick electrodes in it."

Theoretically, if a human ate this fish that hosts this bacterium, we would become a host too. But wanting to stay in us permanently is another matter. Perhaps it would not care to live in something so big.[2]

Why are we getting taller?

It is true that humans have been growing taller throughout history. The increase in the average height of males has been about four inches over the past two hundred years. Female height has been between 6 and 9 percent below that of males.

This phenomenon appears to have occurred in virtually every country in Europe and in the United States, Canada, Australia, New Zealand, and throughout the other developed nations. The increase in height has been associated with industrialization, improved living conditions and nourishment, better medical care, and improved hygiene and sanitation. At least four theories have been put forward to explain our growth:

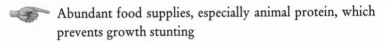 Abundant food supplies, especially animal protein, which prevents growth stunting

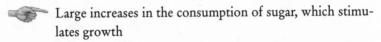 Large increases in the consumption of sugar, which stimulates growth

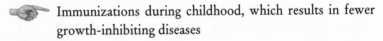 Immunizations during childhood, which results in fewer growth-inhibiting diseases

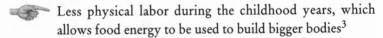 Less physical labor during the childhood years, which allows food energy to be used to build bigger bodies[3]

How small can a person be?

We are capable of doing things in small ways too. The world's smallest human ever was Lucia Zarate (1863–89) of San Carlos, Mexico. She weighed just thirteen pounds at the age of twenty. That was the heaviest weight she ever achieved. At birth, she weighed only 2.4 pounds, and at age seventeen she weighed only 4.6 pounds. Her adult height was only twenty-six inches. Suffering from a particularly devastating form of dwarfism (nanosomia) that is now more commonly called dysplasia, she was unable to grow larger and put on weight.

There are at least twenty-nine separate forms of dysplasia. One of the more common is achondroplasia, which occurs in about one in every twenty-five thousand births. The basis of this condition is genetic, but its cause is unknown.

Ateliosis is perhaps the most growth-restricting condition. Individuals with ateliosis are usually the shortest people of all. They have essentially normal proportions but suffer from growth-hormone deficiency.

Ms. Zarate lived in poverty and probably did not get adequate nutrition. Although genes and hormones are chiefly involved in dysplasias, poor nutrition contributes to such conditions. Poorer nutrition in historical times meant that individuals tended to be shorter.

SAM STACEY, AN eighteen-year-old from Stainforth, England, is reputedly the female with the longest pair of legs. Her legs measure 49.75 inches from hip to heel in January 2001, the height of an average ten-year-old English child.

Although Ms. Zarate's body was tiny, her intellectual capacity was evidently normal. A major myth about people with dysplasias is that a small body means a small brain. In fact, however, although mental retardation does occur in some dysplasias, in most cases the brain and mind are fully capable of whatever "normal" humans can do.

The late actor Michael Dunn is a case in point. "Proud to be a dwarf," he used to say, and he was known to have the finest memory for movie lines in Hollywood. He supposedly could memorize an entire Shakespeare play in one reading. Even though there were few roles for him, he received an Academy Award nomination for his role in *Ship of Fools* (1965) and starred on Broadway. A

DWARFS AND midgets almost always have normal-size children even if both parents are dwarfs or midgets.

MANY KNOW that Anne Boleyn, Henry VIII's second wife, had six fingers on one hand. She used to wear special gloves to conceal that fact. Even fewer know that she also had three breasts.

member of Mensa, an international organization of people with unusually high intelligence, Dunn at one stage held the Mensa performance record.

The world's smallest surviving human baby was born to Marian Taggart (1938–83) in South Shields, England, on 6 December 1961, six weeks premature. Astonishingly, the birth was unattended. The baby girl weighed only ten ounces at birth and was only 1.1 inches long. A normal, full-term newborn weighs about seven pounds and is about thirteen inches long.

About 5 percent of human newborns weigh nearly nine pounds, while about 5 percent weigh less than five and a half pounds. Infants weighing less than that are classified as "low birthweight," and those weighing under 3.3 pounds are "very low birthweight." Even an average twenty-week-old fetus weighs more than half a pound.

When Mrs. Taggart finally received medical treatment for her baby, that attention was nothing short of heroic. Dr. D. A. Shearer fed the tiny infant hourly for the first thirty hours with a solution of glucose, brandy, and water—through a fountain-pen filler. The careful, around-the-clock attention paid off. At three weeks, the tiny girl weighed nearly 2 pounds. At one year, she weighed 2.8 pounds. By her twenty-first birthday, she weighed a normal 105 pounds.

The smallest part of the human body is the human cell. Human cells come in various sizes. The male sex cell, the sperm, is the smallest, in contrast to the female sex cell, the ovum, which is the largest and the only cell capable of being seen with the naked eye.

The stapes, or stirrup bone, is the smallest human bone. It is one of three auditory ossicle bones in the middle ear, weighs 0.000113935 ounces (3.23 milligrams), and measures about one-tenth of an inch. This is in contrast to the largest bone, the femur (thighbone), which accounts for about 27.5 percent of an adult's height.

As for muscles, the smallest is the stapedius muscle, which controls the stapes and is only one-twentieth of an inch long. The stapedius could not be more different in form and function from the body's largest muscle, the gluteus maximus (the muscle of the buttock). In this case, the biggest is always behind.[4]

Can some people predict the weather from the pain in their joints?

It is possible for people to predict the weather by "tuning in" to their pain. According to Dr. Paul Ort, an orthopedic surgeon at New York University Medical Center, people who have had joint injuries may indeed suffer extra aches and pains when the barometric pressure falls, but the medical and behavioral sciences do not presently know why.[5]

Dr. Joseph Hollander, of the University of Pennsylvania Medical School in Philadelphia, says that a rise in humidity and a decrease in barometric pressure often leads to an increase in pain for those suffering from arthritis. Cell permeability may help explain the pain.

More blood fluid may be forced into the tissues of arthritis sufferers. Blood-vessel walls are often more permeable in people who have arthritis, and blood is always under higher pressure than the surrounding body tissues. This movement would be greatest when the pressure of the surrounding environment outside the body is lowest—as it is just before a storm. If joints are already sore, stiff, swollen, and inflamed, the added fluid could trigger the extra pain. This still unproven theory is perhaps the best that science has come up with so far.

Dr. Melvin Rosenwater, an orthopedic surgeon formerly at the Columbia College of Physicians and Surgeons in New York and now living in Boca Raton, Florida, is convinced that nerve cells in joints may be sensitive to changes in barometric pressure. He calls this "the barometric ache" but hastens to add, "It is not a real clinical problem."[6] He stresses that barometric ache may be treated with anti-inflammatory medications, such as aspirin, or with a trip to a different climate.[7]

Can our bodies warn us of earthquakes?

Throughout history, people have occasionally claimed to have this ability. Although it has never been historically documented, a Native American by the name of Shotola supposedly warned the famous Italian opera singer Enrico Caruso not to remain in San Francisco—just before the 1906 earthquake and fire devastated the city. Shotola claimed that he could "hear earthquakes" before they happened.

Today after a major earthquake strikes, there are often some who boast that they knew it was coming. However, people who come forward only after the event are rarely believed. The public generally, and scientists in particular, are rightly dismissive of such claims. In fact, history fails to record a single case where an individual truly accurately predicted an earthquake using only their senses. Thus, it seems that the human body cannot warn us of earthquakes.

Nevertheless, there is striking evidence that animals may possess some form of an earthquake-predicting ability. Dr. Helmut Tributsch, a German physicist and chemist, argues that animals unquestionably have the ability to predict earthquakes and that the ancient Greeks were the first to recognize this in animals.[8]

Dr. Tributsch maintains that hours, even days, before earthquakes, distinct changes in the behaviors of animals can forecast that an earthquake is imminent—long before the sophisticated seismographic equipment can. Dr. Tributsch gives some examples of changes in animal behavior immediately before an earthquake:

 Animals exhibit unusual behaviors: Animal unrest occurred in a Peruvian town just before a major earthquake: Chickens would not return to their coops, mother cats carried their litters out of homes and into open spaces, cows mooed eerily and incessantly, and so on.

A PHYSICIAN who treats the hands and feet is a chiropodist.

AN ECTRO-dactylyte is the name for people who have lobster-like hands or feet. The most famous was John Merrick, "the Elephant Man."

Fish change their migratory patterns: Fishermen in Japan have long recognized changes in the migratory patterns of fish right before an earthquake. A 1932 statistical report on mackerel catches near Japan's Izu Peninsula seems to confirm this: The annual size of the fish haul during the 1920s varied directly in relation to earthquake activity. Furthermore, the shallow coastal waters near Miyagi, Japan, were teeming with schools of sardines, which are rare in the Pacific, before a major earthquake struck near there in 1933. In addition, Japanese folklore claims that fish foretell earthquakes. Before the Tokyo earthquake of 1855, a fisherman noted a large number of catfish frantically splashing on the surface of the water. Recalling this point of folklore and realizing that catfish are normally sluggish, bottom-dwelling fish, the quick-thinking fisherman was able to return home and drag the furniture out of his house just before the earthquake struck.

 Rats migrate: A restaurant called the House of Rats in Nagoya, Japan, lost its claim to fame when its rats, which had been allowed to roam free in the restaurant, suddenly disappeared. On 27 October 1991, one day after the rats left, an earthquake struck with an estimated force of 7.9 on the Richter scale. And the day before a major earthquake struck the San Fernando Valley on 9 February 1971, police in California reported large numbers of rats scurrying through the streets.

 Snakes migrate: The 7.3 earthquake that struck the Liaoning Province of China on 4 February 1975 and caused massive destruction should have caused more loss of life, but few people were killed because most had already evacuated the area. They did this because they had noticed unusual animal behaviors that began as early as two months before the earthquake. Most notable was that snakes had come out of winter hibernation to migrate but instead were freezing to death in the snow.

THE MOST sensitive finger on the human hand is the index finger.

THE FEMUR (thighbone) is the strongest bone in the human body.

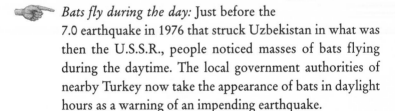 *Bats fly during the day:* Just before the 7.0 earthquake in 1976 that struck Uzbekistan in what was then the U.S.S.R., people noticed masses of bats flying during the daytime. The local government authorities of nearby Turkey now take the appearance of bats in daylight hours as a warning of an impending earthquake.

Birds do not roost: Before the major earthquake hit Friuli, Italy, in May 1976, local residents reported seeing many odd animal behaviors, including birds emitting strange calls and nesting birds refusing to roost.

Cattle head for high ground: In March 1964, a cattleman on Kodiak Island, Alaska, decided to head for higher ground, following a hunch based on the behavior of his cattle. After herding his cattle the same way for years, the cattle one day suddenly stopped grazing and just as suddenly started heading for higher ground. Puzzled, the cattleman followed the cattle. Two hours later, a major earthquake struck, devastating the city of Anchorage. This was followed by a huge tidal wave that flooded the lowlands that the cattle and the cattleman had just left.

 Chimpanzees become hyperactive: At the Primate Study Center of Stanford University, researchers routinely track the behavior of chimpanzees. In 1975, they noted that an earthquake was far more likely to occur on the day after a day in which the chimps were the most active. In addition, after the October 1989 Loma Prieta earthquake near San Francisco there were anecdotal reports of strange animal behavior taking place before that quake. For example, "at the aquarium in Monterey (south of the epicenter), two divers in a tank noticed that fish stopped swimming in formation just before the quake—and returned to formation afterward."[9]

Dr. William Bakun, chief seismologist for the U.S. Geological Survey in Menlo Park, California, admits that there are "ample reports world-wide of animals going haywire prior to earthquakes."[10]

Why does animal behavior change?

As to what about earthquakes alters the behavior of animals, Dr. Tributsch speculates that it is probably a combination of factors: "Venting ground gasses could excite animals . . . [or] perhaps changes in the groundwater table." Moreover, animals may sense "changes in atmospheric pressure, slight swaying of the ground—even ultrasound emitted from rock that is nearly bursting."

Dr. James Berkland, a geologist in Santa Clara, California, contends that changes in the earth's magnetic field, which he believes occur before an earthquake takes place, "disorient" animals and change their behavior. This controversial view is frowned on by many scientists, but it has recently been bolstered by what is perhaps a coincidental event. According to Dr. Anthony Fraser-Smith, an atmospheric researcher at Stanford University, the Stanford magnetometer recorded "highly unusual changes" just before the Loma Prieta earthquake. The magnetometer usually detects only magnetic-field variations due to solar magnetic storms and other effects high in the earth's atmosphere.

Could we ever rely on animals for earthquake prediction? Dr. Bakum does not dismiss the possibility. He claims that perhaps "if we understood what these creatures were responding to, we could build an instrument to measure that phenomenon."

Dr. Berkland has a simpler strategy. Because one of the pre-earthquake behaviors is that more pets run away from home, he thinks that scientists should monitor the lost-and-found ads in newspapers.

If animals can predict earthquakes, could humans ever learn such a skill from animals? Theoretically, the possibility exists, but it seems highly improbable. Still, if someone comes to you and says that he hears earthquakes—it doesn't hurt to listen.[11]

THERE ARE twenty-six bones in the human foot.

ABOUT 20 percent of your body mass is the skeleton.

THE HARDEST bone in the human body is the jawbone.

What crippled Tiny Tim?

In 1972, Australian researchers seemed to have solved this recurring Christmas medical mystery: What was wrong with Tiny Tim? Now, we're not so sure.

Tiny Tim, the Cratchit family's crippled little boy in *A Christmas Carol* (1843), is one of the most beloved characters created by Charles Dickens (1812–70). It is often assumed that Tiny Tim suffered from polio, but that was probably not the case.

Tiny Tim's condition was the subject of an insightful and imaginative article appearing in the *Australian Paediatric Journal*.[12] Its author, Dr. Peter Jones, is now retired from his post at the Royal Children's Hospital in Melbourne. According to Dr. Jones, there is only "sparse information" on which to base a reasonable diagnosis. We know from Dickens's description only that Tiny Tim "bore a little crutch and had his limbs supported by an iron frame." Furthermore, he was carried around on the back of his father, Bob Cratchit.

Although this sounds very much like a child with polio, Dr. Jones points out: "Tuberculosis of the hip is a far more likely diagnosis; 'coxalgia,' as it was known from its presenting symptom, carried with it a sentence of gross deformity and frequently death." Polio, he says, is an unlikely diagnosis because "poliomyelitis is paradoxically more prevalent in communities with high standards of municipal and domestic

HUMANS ARE born with three hundred bones in the body. But by adulthood, we have only 206 bones, because many of our original bones fuse together.

THE SENSITIVITY of a woman's middle finger is reduced during menstruation.

hygiene—and no one would suggest that this could be said of London in 1843."

As for Tiny Tim's chances of recovery, Dr. Jones prefers to "carry optimism to the extreme as Dickens did" and suggest that Tiny Tim would survive. He adds that Tiny Tim might have had "pseudocoxalgia" or Perthes' disease, a form of osteochondrosis (although this disease was not defined for another seventy-seven years).

However, a U.S. doctor claims that Tiny Tim did not suffer from a bone disease at all, but instead had a kidney disease that made his blood extremely acidic and eventually resulted in crippling him.[13]

Dr. Donald Lewis, a professor of pediatrics and neurology at the Medical College of Hampton Roads in Norfolk, Virginia, studies Tiny Tim to show medical students how to diagnose children. Dr. Lewis has his students read Dickens and search for symptoms.

According to Dr. Lewis, Dickens could not have known about the kidney condition called DRTA-1 (Distal Renal Tubular Acidosis Type 1) because the disease was not recognized until the early twentieth century. Nevertheless, therapies for symptoms of the disease were used in Dickens's day and might have successfully treated it. Dr. Lewis argues that Tiny Tim had classic symptoms of DRTA-1, but also may have suffered from neurological problems, as evidenced by Tiny Tim's spells of weakness, his withered hand, and his limpness.

In any case, the impact of Dickens's *Christmas Carol* on nineteenth-century England was so great that a charitable trust for crippled children was founded—the Tiny Tim Guild.

God bless us, every one![14]

Why do my feet swell in airplanes?

This OBQ comes up during many radio interviews, perhaps because of the recent publicity about DVT (deep vein thrombosis) on long aircraft journeys.

It is a myth that the feet of airplane passengers swell because of changes in atmospheric pressure due to high elevation. Feet swell on planes, especially during long flights, for the same reason they swell on the ground—inactivity.

The heart is not the only body organ that serves as a pump—leg muscles do too. Walking or flexing a leg assists in the pumping effect of the heart. If, while on an airplane, you are not only confined but also sitting, gravity forces blood and other fluids to the lowest body point: the feet.

The pooling of body fluids in the feet can happen just as easily in a bus, on a train, and in a car or office. In fact, most people's feet normally swell somewhat during the day anyway—some up to a full size larger.

If you remain inactive while on a long flight, it does not matter whether you leave your shoes on or off. They will swell either way. If left on, shoes will provide external support but will inhibit circulation a bit more and probably feel tighter during the latter part of the flight. If shoes are removed, comfort is probably increased, but the shoes are likely to be more difficult to put on once the flight is over.

Podiatrists normally recommend airplane aerobics to help circulation, and to minimize swelling.

THE KNEE IS the most easily injured joint of the human body. In U.S. hospitals each year, 1.4 million patients are admitted with knee problems.

Can a person be identified by DNA from a bit of bone?

It is true that scientists can now identify a missing person from just a tiny fragment of bone, a single tooth, or a bit of DNA. Skeletal remains, bones and teeth, can reveal the age and height at death, as well as sex and even ethnicity.

According to Drs. Christopher Joyce and Eric Stover, forensic anthropologists have some useful keys to unlock the identification door.[15]

 Age: When scientists examine human skeletal remains to determine the age of the person at the time of death, they first look at the back of the skull. Although the occipital bone at the base of the skull fuses and hardens at age four or five, eight other bones in the head continue to grow as a person gets older. This means that together the condition of these bones can establish an age range of up to forty years. Next, scientists look at the teeth. The pattern of teeth appearance (eruption) is also useful in determining age. For example, the dental remains may consist of only deciduous (baby) teeth, only permanent teeth, or a combination of both.

 Sex: When the gender of a person is sought, scientists begin by examining the pelvic region. A woman's pelvic bone opening is wider than a man's, to better accommodate childbirth. Next, the front of the skull is examined. In men, the ridge of the brow is almost always more prominent than in women.

☞ *Height:* When height is sought, scientists first look to the thigh. The length of the femur (the long bone of the thigh) is directly related to a person's overall height. If the femur is available, or even just enough of it so that its length can be calculated, the entire height of the person can also be calculated. For this last step, forensic anthropologists use special, anthropometric tables.

☞ *Ethnicity:* When scientists want to determine ethnicity, they again look to the skull. Forensic anthropologists claim to be able to successfully identify ethnicity in about 90 percent of cases by measuring a number of subtle skull differences.

☞ *Medical and dental records:* Although skeletal remains themselves give forensic anthropologists valuable evidence, skeletal data in combination with medical or dental records yield even greater information. When a body needs to be identified, scientists make observations and conduct tests of the skeletal remains, then compare observations and test results with the medical or dental records of a "suspect."

In addition to examining remains for age, sex, height, and ethnicity, there are three other means of establishing identity when both skeletal evidence and medical records are available.

First, dental work documented in dental records—number of teeth, bridgework, and other aspects of dental composition—may by itself lead to a positive identification. Teeth are coated with tough enamel, so they are more resistant to deterioration than other parts of a skeleton. Furthermore, each tooth is unique in the shape of its crown, roots, and pulp cavity.

Second, injuries to bones or teeth often leave permanent marks that can be documented in medical or dental records.

Third, as with teeth, bones are unique in their size, shape, and the details of their internal structure. These elements are readily revealed by X-rays, which can then be compared with those in the medical records. According to Dr. Clyde Snow, a University of Oklahoma forensic anthropologist, "bones are just as unique as fingerprints," which makes it possible to identify a person from a fragment of bone or even just one tooth.[16]

But what if no medical or dental records are available, or there are only a few fragments of bone? Advances in research now make it possible to extract genetic material—DNA—from bone and teeth fragments of people, even if long dead. Identity can be established by comparing these samples with samples from living relatives. In this process, a particular DNA—mitochondrial DNA—is especially useful.

In every human cell there are structures shaped like coffee beans called mitochondria. These are located outside the cell nucleus but well within the cell wall. The mitochondria generate energy for the cell by breaking down sugars and fats. Within the mitochondria, DNA is found in great abundance, often in amounts hundreds of times greater than in other parts of the cell. Experience has shown that if any DNA survives it is most likely to be the DNA of the mitochondria. Because

THE AVERAGE person takes eight thousand to ten thousand steps a day and may walk up to 115,000 miles in a lifetime. That is the equivalent of five times around the equator.

THE THUMB IS such an important part of the human body that it has a special section in the brain, separate from the section that controls the fingers, reserved for regulation and control of the thumb.

A PERSON afflicted with hexadactylism has six fingers or six toes on one or both hands and feet.

mitochondrial DNA is passed only from a mother to all her offspring, a person's mitochondrial DNA is identical to that of his or her mother, sisters, or brothers. Barring a rare genetic mutation, identifying a person this way is nearly always possible.

Thus, in order to establish identity, scientists move up and down a family's tree, using DNA as the ladder.[17]

Why are pygmies so short?

African pygmies have fascinated Westerners since the first photographs of these miniature people reached Europe more than a century ago. But pygmies have also baffled scientists, because the normal explanation for short stature simply does not apply to pygmies.

Short stature is usually caused by low levels of human growth hormone (HGH). This hormone is secreted during the night by the pituitary gland, located in the brain. Shortness (or tallness) runs in families because the amount of HGH an individual possesses is genetically determined. The gene for HGH is one of a cluster of five closely related genes located on chromosome 17. The hormone itself is actually a complicated protein made up of 191 amino acids.

Growth will be reduced if the pituitary gland fails to secrete enough HGH, if a person is afflicted with a genetic abnormality involving chromosome 17, or if HGH is adequate but for some reason cannot be put to use by the body.

Growth can also be stunted by nutritional deficiency, illness, or injury, evidence of which can be seen throughout the starving Third World.

Any abnormally undersize person is medically termed a "dwarf." Dwarfs exhibit a condition known as nanosomia, sometimes called nanism. For years an artificial, genetically engineered, biosynthetic growth hormone has been available in the United States and Australia for treating nanosomia. According to Dr. G. L. Warne of the Royal Children's Hospital in Melbourne, biosynthetic HGH "sounds like a miracle drug, and in many ways it is. It has so far not been associated with any serious side effects." However, Dr. Warne points out, "biosynthetic HGH is extremely expensive, costing $20 a unit, and the cost of a year's treatment for a 30 kilogram child [about 66 pounds] is at least $15,600. The course of treatment, involving daily subcutaneous injections, can last for many years."[18]

For nearly forty years, scientists have known that pygmies have normal amounts of HGH, And, compared with their taller African neighbors, pygmies have an adequate diet, enjoy reasonable health and life expectancy, and do not experience abnormal rates of injury. So why is it that pygmies rarely grow taller than four feet, six inches?

The most recent scientific thinking suggests that pygmies are short because of a "cell receptor failure" in their bodies, meaning that although they have sufficient amounts of HGH, they fail to process it correctly.

A cell receptor normally provides an all-important "docking site" for the HGH circulating in the bloodstream. The HGH must "dock" with a cell in order to be utilized by that cell. But in pygmies the cell receptor is deficient and does not allow enough docking to go on. Hence, this cell receptor failure causes much of the HGH to be wasted.

Drs. Gerhard Baumann and Melissa Shaw of Northwestern University Medical School in Chicago, along with Dr. Thomas Merimee of the University of Florida in Gainesville, analyzed the blood of twenty pygmies living in the Ituri Forest of the central region of the Democratic Republic of the Congo (formerly Zaire). They came to some interesting conclusions.[19]

> IN THE FUTURE IT will be possible for people to regrow missing arms or legs, just like a salamander can grow a new tail. Research has shown promising results in getting bone to grow with the application of electricity. Already, children under the age of five who have lost the tip of a finger achieve complete regrowth.

The Baumann team found "debris resulting from cell receptor failure" within pygmy blood. This debris consisted of molecules of HGH attached to a binding protein broken off during an unsuccessful docking attempt. The team also discovered that pygmies have half the amount of the necessary hormone-protein complex, compared with a group of individuals serving as control subjects. Thus, this low level of the hormone-protein complex "strongly indicates" that pygmies do in fact have a shortage of HGH cell receptors. With so few receptors, making docking difficult, pygmies almost certainly would have problems processing HGH—even if they received it artificially.

The theory put forward by the Baumann team is consistent with the often-observed characteristics of pygmy growth. Pygmies show seriously stunted growth during adolescence, which is precisely the time when a full complement of cell receptors is most essential for normal growth. In fact, normal Western teenagers usually produce extra HGH cell receptors during this period, thus allowing the typical adolescent growth spurt.

The Ituri Forest pygmies demonstrate that although they are short they are certainly extremely able-bodied. Both men and women are exceedingly agile, nimble, and strong. But their children pose perhaps the greatest fascination. According to Jean Pierre Hallett in both the book *Pygmy Kitabu* (1973) and the film *Pygmies* (1974), pygmy children from a very young age have extraordinary physical abilities.[20] For example, three-year-old pygmies can accurately shoot an arrow, hitting a two-and-a-half-inch-wide target nine out of ten times at a distance of thirty feet. They can use a sling to down small prey, swim confidently, and run faster than the average Western child at age eight. At age three, the typical Western child is scarcely able to balance on one foot, while a three-year-old pygmy child can effortlessly scale a sixty-foot-high coconut tree.

Actions speak louder than size.[21, 22]

Do we really have a "funny bone"?

We do not really have a "funny bone"—it is really a "funny nerve." The nerve is the ulnar nerve, which supplies sensation to the arm, hand, and fingers. For most of its length, the ulnar nerve is located deep under the skin, where it is well protected.[23] However, at the elbow it comes very close to the surface and is covered only by skin and a thin layer of connective tissue.

That is why it hurts so strangely when you bang your elbow in a certain way. You have actually momentarily traumatized your ulnar nerve. The pain sensation is over in a few seconds.

It's ironic that someone named it "funny."[24]

What is writer's cramp?

Writer's cramp is actually a localized muscle spasm called focal dystonia, caused by holding a pen or pencil too long, especially too tightly. Relaxing the hand periodically, exercising the hand, holding the pen more loosely, and taking frequent breaks from writing usually solve the problem. To test whether you are holding the pen loosely enough, see if your nonwriting hand can easily pull the pen from your writing hand. A physician or other health professional would recommend exercises and relaxation techniques that help overcome writer's cramp. However, if a serious injury results, surgery is done—but only in the most severe cases.[25]

Drs. Lee Tempel and Joel Perlmutter from Washington University in St. Louis studied blood flow in patients with writer's cramp and compared it with the blood flow of people without such symptoms. They found that, on average, the patients with writer's cramp showed only two-thirds as much blood flow in the brain's sensorimotor cortex, the region of the brain responsible for hand sensation and movement.[26, 27]

What causes a "side ache" in my ribs?

The pain of a side ache is really not in your ribs. A "side ache," "runner's cramp," or "stitch," as it is variously called, is indeed a cramp. A cramp is a painful spasmodic muscular contraction, especially a tonic (tension) spasm.

But a side ache actually occurs in the intestines. Dr. Averil Ma, a gastroenterologist at the Columbia Presbyterian Medical Center in

New York, explains: "Several things can cause such a cramp. For example, if someone ate a heavy meal and then started doing something else that demands blood flow, there would be relative ischemia (a reduction in the blood supply to the intestines). This would produce a cramp. Muscular problems can also cause such pain."[28]

THE LENGTH from your wrist to your elbow is the same as the length of your foot.

THE FEET account for one-quarter of all the bones in the human body.

Do people who have lost a limb still sometimes feel sensation in it?

This refers to the fascinating phenomenon of Phantom Limb Syndrome (PLS).

Sensation in lost arms and legs reported by amputees has intrigued physicians throughout the centuries, and PLS has been studied and reported in the medical and behavioral science literature since at least 1940.[29]

According to Dr. Ronald Melzack of the Department of Psychology at McGill University in Montreal, PLS occurs in up to 70 percent of amputees. The sensation is often painful and has been described as burning, cramping, or shooting. It can vary from occasional and mild, to continuous and severed. It usually starts soon after the amputation but sometimes appears weeks, months, or even years later.[30, 31]

Dr. Melzack adds: "The oldest explanation for phantom limbs and their associated pain is that the remaining nerves in the stump, which grow at the cut end into nodules called neuromas, continue to generate impulses."

Why do people shrink when they get older?

Seniors do not shrink as such, but they do experience a loss in body height. What happens is due less to genes and more to the effects of gravity and time. By age seventy, the average man of average weight and height loses about one and a quarter inches from his maximum height achieved during early adulthood.

According to Dr. Lawrence Riggs of the Mayo Clinic in Minneapolis, three factors are involved. First, there is a loss of space between the spinal discs due to the pressure of gravity. These small gaps have a cumulative effect, given the large number of discs. Second, with age comes a general weakening of the back muscles. Third, as we age we usually have experienced many years of poor posture. Women suffer a slightly greater loss in proportion to their earlier height. This is attributed to the higher rates of osteoporosis in women.[32]

Why am I taller in the morning?

All of us are taller in the morning, shorter in the afternoon, and shortest of all at night.

Dr. Jerry Wales of the Department of Paediatrics at the University of Sheffield explains that there are two components in this. First, he says, in the growing child "growth hormone is secreted in pulses overnight. This acts through several intermediary steps to cause lengthening of the bones at the end-plates (epiphyses)." Second, after growth ceases, there is a daily "postural compression of the spine under the effect of gravity." The difference in adults is an average of roughly 0.59 inch from morning to night.

Dr. Peter Dangerfield of the Department of Anatomy at the University of Liverpool adds that another component must be considered: the existing curves of the spinal column itself. "These curves," he says, "vary with body weight and position. As a result, the spinal column tends to press downwards when in an upright position, altering these curvatures, and hence shortening the spinal length. When lying down, the reverse happens and the column lengthens again. It is estimated that 80 percent of the height change is accounted for by changes in these curvatures."[33]

chapter 9

the **inside**

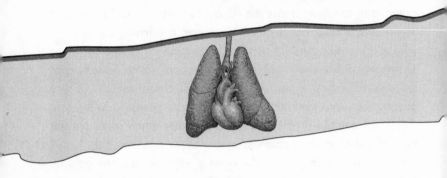

The ancient Greek playwright Sophocles (495–406 B.C.) once wrote, "Wonders are many, and none is more wonderful than man." Yes, humans are wonderful. The deeper we probe into this wonderful organism, the more we reveal how wonderful we are. So let's look inside.

What is pain?

We all feel pain, but what is pain, really? Pain is defined medically as a more or less localized sensation of discomfort, distress, or agony resulting from the stimulation of specialized nerve endings. But research shows that pain is far more complicated.

Pain is necessary for our survival. It serves as a protective mechanism—the body's danger alarm—insofar as it warns us to remove ourselves, if we can, from the possibly hazardous or even fatal pain source. For example, when we burn our fingers with a match, the body is telling us: "Do not expose me to this high temperature any longer or permanent cell damage will result."

Medical science distinguishes at least thirty-six separate kinds of pain. "Bearing-down pain" accompanies uterine contractions during the second stage of childbirth labor. "Growing pains" are recurrent quasi-rheumatic limb pains peculiar to adolescence. Sharp, darting pains are called "lancinating pains."

Some pains are bizarre and difficult to explain. For example, "phantom limb pain" is felt as arising from a limb that has been amputated. Some pains are named after people. "Brodie's pain" is induced by folding the skin near a joint affected with neuralgia. "Charcot's pain" results from rheumatism of a testicle.

Research shows that pain is not merely a simple electrical nerve response. Actually, pain is relatively independent of the body's own pathological process. Many studies show that pain may be reduced with a placebo (a dummy medical treatment). It also can be psychologically induced through the use of a nocebo (a dangerous medical treatment). This has been demonstrated many times.[1]

Pain stimuli may also evoke differing behavioral responses according to age, prior psychological status, type of group the

THE AVERAGE adult has between forty billion and fifty billion fat cells.

THE WORLD'S largest chest belonged to Robert Earl Hughes of Illinois. It spanned ten feet, four inches, and he weighed a mammoth 1,067 pounds when he died in 1958 at age thirty-two. He was buried in a coffin the size of a piano crate.

person belongs too (for example, a religious group), social context, and culture.

The amount of pain inferred or attributed to nine different conditions is "culturally learned," according to two University of Alberta psychologists, Drs. Janice and Robert Morse, who studied subjects from four different cultural groups residing in western Canada: Hutterites, East Indians, Ukrainians, and Canadian Anglophones. The study found that each of the four groups rated and reacted differently to each of the nine different pain stimuli.[2]

Thus, pain is your experience speaking too, not just your nerves.[3]

THE UTERUS IS the organ that has given us the word "hysterical."

IF LAID OUT IN A straight line, the average adult's circulatory system would be nearly sixty thousand miles long—enough to circle the earth two and a half times.

What is a laugh?

Physiologically, laughing is a series of spasmodic and partly involuntary expirations with odd vocalizations that are normally indicative of merriment. Often, however, laughing is a hysterical manifestation or a reflex result of tickling. Normal laughing is of two types, mild and heavy, and there is a socially accepted occasion for each. There are three types of abnormal laughing: compulsive, forced, and obsessive.

What makes me laugh?

It is surprising, however, that, other than as the result of simple tickling, laughing is based on fears. These fears include fear of social embarrassment, of loss of dignity, of exclusion from a group, of being fooled or exploited, of death, of injury, of sex. There is a fine line between comedy and tragedy, between what makes us laugh and what makes us cry, between pleasure and pain.

What happens to my body when I laugh?

When you give way to laughter, electrical impulses are triggered in nerves in your brain. These impulses set off chemical reactions in the brain and elsewhere in the body. For example, your endocrine system orders your brain to secrete natural tranquilizers and painkillers. Other released chemicals aid digestion, and still others make arteries contract and relax and improve blood flow.

Laughing may not be the best medicine, but it's certainly a good one.

Why is laughing important?

Among other things, laughing restores balance and equilibrium. Charles Darwin, in *The Expressions of the Emotions in Man and Animals* (1872), argued that laughing helps us discharge surplus tension and mental excitation. Sigmund Freud argued that laughter helps us deal with lustful thoughts.

Laughing is important to our very survival. The ability to laugh appears when at about

THE HUMAN heart beats about 70 times a minute. The shrew's heart beats 600 times a minute, the hummingbird's 1,300 times a minute, and the blue whale's 10 times a minute.

YOUR HEART
will thump
approximately
42,075,900 beats
a year and about
three billion
times in an
average lifetime,
give or take a
few million.

HIPPOCRATES, THE
ancient Greek
"Father of
Medicine,"
believed that a
flat-chested
woman could
enlarge her bust
by singing loudly
and often.

twelve weeks of age. Darwin argued that a baby's laugh gives pleasure to the caretaker and thus helps lessen the likelihood of parental rejection—which aids both personal survival and the survival of the species.

Can laughing keep us healthy and even heal us?

The late Dr. Norm Cousins, in his book *Anatomy of an Illness as Perceived by the Patient* (1979), claimed that laughter helped him regain his health. His evidence was so overwhelming that some hospitals now have "laughing libraries" of videos to help patients laugh their way to recovery. There are also "clown wards" in hospitals—especially children's wards.[4]

It is said that "a laugh a day keeps the doctor away," but is it true that laughter has curative powers? Research shows that when we laugh our metabolism rate picks up, muscles are massaged and stimulated, and a variety of biochemical substances rush into the bloodstream. After a period of laughing, subjects feel momentarily relaxed. Laughter also fortifies us against depression and heart disease and heightens our resistance to pain.

Now U.S. researchers think that laughter may boost the immune system as well. This has been shown in several experiments. For example, in one experiment conducted by Dr. Kathleen Dillon at the Western New England College in Springfield, Massachusetts, university student volunteers were divided into two groups. One group was instructed to watch a nonhumorous educational video. The other group watched a humorous video of Richard Pryor comedy routines. Dr. Dillon found that concentrations of salivary immunoglobulin (IgA), an antibody linked to lower rates of upper respiratory illness, jumped measurably in the "humor video" group.[5]

According to Dr. Lee S. Berk, an immunologist at the Loma Linda University School of Medicine, "negative emotions can manipulate the immune system, and it now seems positive ones can do something similar." Although laughter is the subject of much speculation, Dr. Berk believes that laughter starts a simple biochemical process in motion involving the body-produced hormone cortisol: "Cortisol, which is an immune suppressor, has a tremendous influence on the immune system. Laughter decreases cortisol, which allows interleukin-2 and other immune boosters to express themselves."

Just as the Robin Williams film *Patch Adams* (1998) portrayed, the possibility of the curative powers of laughter has already begun to motivate change in the practice of some hospitals. For example, on an experimental basis, Columbia Presbyterian Medical Center in New York City has established "The Big Apple Circus-Clown Care Unit," where laughter is continually emphasized. Professional

A POLYORCHID man has at least three testicles.

THE LIVER IS the largest gland in the human body.

THE HUMAN heart creates enough pressure when it pumps blood out of the body to squirt blood thirty feet.

clowns dress in white hospital coats and race around the wards on roller skates. Practical jokes and pranks are constant. "Let's draw your blood," says a "staff" member, and a crayon and sketchpad are produced. So far, jokes that bomb have not resulted in malpractice suits.

No one knows for sure whether "a laugh a day keeps the doctor away," but it gives new meaning to a Stephen Sondheim song. Illness may be the time to "send in the clowns."[6]

Does laughing make us more productive?

Research shows that industrial productivity is boosted by humor, and evidence suggests that humor in the workplace is a powerful profit-making tool. Some of the world's largest corporations, such as IBM, Monsanto, and General Foods, are cashing in.

Dr. Alice Isen's studies at the University of Maryland show that "people put in a good mood organize information better and are more creative," that "people in good spirits proved more creative in word association, categorization, and tasks involving memory," and that "humor also improves decision making and negotiating abilities."[7]

Dr. Isen says that "mild elation seems to lead to the kind of thinking that enables people to solve problems requiring ingenuity or innovation," adding, "Someone who is happy can perceive subtle relationships between things because positive material is stimulated in his memory. He has more ideas."

One of Dr. Isen's experiments produced some intriguing results. Adult volunteers were divided into pairs playing fictitious roles of buyers and sellers of goods. They had to successfully perceive a range of alternatives and combine them cleverly in order to achieve the highest profits or the best deal. It is interesting that the pairs who viewed humorous cartoons beforehand were less contentious and "likelier to reach solutions that were mutually beneficial."

Dr. Isen says that "combining issues and developing novel solutions may be necessary for anything that goes beyond obvious compromises. These are the very same capabilities that are enhanced by positive feelings."

At the University of Tennessee, Dr. Howard Pollio's work shows that "humor improves group as well as individual performance." In fact, "when it's related to the task at hand, laughter tends to boost performance." Dr. Pollio asserts that laughter provides a "brief break without being too diverting," relieves boredom, and encourages problem-solving.

At California State University, Long Beach, Dr. David Abramis surveyed 341 adults and found that they "take fun at work very seriously" and consider "a sense of accomplishment to be fun." He discovered a "clear relationship between intending to have fun and actually having some: Those who think fun belongs at work enjoy themselves most." So the humor research message is clear: "Lighten up a little and laugh it up a lot. It can only do you—and your employer—a whirl of good."

IT'S ESTIMATED that it would take 1.12 million mosquito bites to drain all the blood from the average adult human being.

THE LARGEST vein in the human body is the inferior vena cava, the vein that returns blood from the bottom half of your body back to your heart. You have every right to ask: But if it is the largest vein why is it called "inferior"?

Related research indicates that sadness, lack of humor, and pessimism may detract from employment performance and may even negatively affect a person's health and life expectancy.

A monumental study spanning thirty-five years followed graduates of Harvard University through their careers. Dr. Christopher Peterson of the University of Michigan, and two colleagues, reported in the *Journal of Personality and Social Psychology* on this longitudinal study of the attitudes and health of the elite.[8]

The Peterson team found a clear relationship between an optimistic outlook and success, and that, correspondingly, there is a relationship between a pessimistic outlook and lack of success in career accomplishment. Even more profound is that "individuals who explain bad events pessimistically in early adulthood (at the time they were first surveyed at Harvard as recent graduates just after age twenty-three) have substantially more illness at age forty-five than those who offer rosier explanations for bad events. The relationship between pessimism and poor health declines somewhat in the following years but remains statistically significant through age sixty."

A follow-up study is being considered by Dr. Peterson. It has been suggested that a twenty-year follow-up to be published in 2008 would be ideal.[9]

Such "good humor" research is being taken very seriously by corporations. The personnel director of the New York office of one multinational says, "We now try to hire MBA grads who are 'up' as well as 'bright.' Who needs execs who

EVERY SECOND, the body's bone marrow creates three million blood cells. During that same second, it destroys the same number.

THE RIGHT LUNG takes in more air than the left lung.

get pensioned off at fifty? I suppose you *could* say that we want the best, the brightest, and the funniest."

There are more than two hundred articles in the medical literature attesting to the fact that having a good sense of humor is a factor in getting well and staying that way.[10]

Does laughing make me cope better?

Is simply having a good laugh the best way to cope with tough times? Perhaps this *is* so for most of us. But how far can you take the notion that humor fights depression? Research has established that humor has important educational and psychological uses as well.

Humor can be used as an educational tool to increase rates of learning. This was demonstrated in two famous experiments by Dr. Avner Ziv of Tel Aviv University published in the *Journal of Experimental Education*.[11] In the first experiment, Dr. Ziv divided 161 university students doing a one-semester statistics course into two groups. One group was taught statistics with humorous teaching materials. The second group was taught the identical subject matter with traditional, nonhumorous materials. At the end of the semester, both groups were tested to see which group, if any, learned the subject matter better. Dr. Ziv writes, "The results showed significant

AT ANY GIVEN moment, twenty-five *trillion* cells travel through the bloodstream.

A STACK OF FIVE hundred blood cells would measure only one twenty-fifth of an inch high.

FOR REASONS unknown, women's bodies reject a heart transplant more often than men's bodies do.

YOUR HEART will beat faster during a brisk walk or during a heated argument than it will during sexual intercourse.

DURING YOUR lifetime, you'll eat about 66,140 pounds of food, the equivalent of the weight of six elephants.

A NEW LAYER of mucus is produced by the stomach every two weeks so that stomach acid will not devour the stomach lining.

differences between the two groups in favor of the group learning with humor."

Humor can also be used as a therapeutic tool in clinical psychology in several ways. Summarizing these uses in an article in *Psychological Reports,* Dr. Sharon Dimmer from the Department of Clinical Psychology at Michigan State University, along with two colleagues, writes: "Humor in psychotherapy can be used to alleviate anxiety and tension, encourage insight, increase motivation, create an atmosphere of closeness and equality between therapist and client, expose absurd beliefs, develop a sense of proportion to one's importance in life situations and facilitate emotional catharsis."[12]

But despite these beneficial uses, is humor an effective weapon against depression? On this, authorities are not so sure.

On the one hand, we all know that laughing makes us feel good—at least for the moment. Furthermore, many psychology textbooks and clinical therapeutic handbooks recommend the use of humor and laughter in both the diagnosis and treatment of psychological disorders, including depression. For example, in the *Handbook of Humor in Psychotherapy,* edited by Drs. William Fry and W. Salameh, it is claimed that humor followed by laughter is incompatible with depression and thus together comprise an extremely valuable treatment against depression.[13, 14]

What comes first in laughing: the smile on the outside or the pleasant feeling on the inside?

M ost people, including many behavioral scientists, would confidently say that the feeling produces the smile, not vice versa. But if the latest research findings are right, then those people are in for a surprise.

Laboratory experiments suggest that the physiology of smiles and of other facial expressions may itself cause emotions. This does not mean that facial expressions are *more* important than thoughts or memories in prompting emotions, only that both interact in ways we previously thought improbable.

Actually, this idea is not really new. More than a century ago, Charles Darwin and psychologist William James (1842–1910) both put forward the theory that facial expressions play an important role in bringing about the feelings that accompany those expressions. "Wipe that look off your face," said Darwin, "and you mute the inner feeling." Over the years, however, the

IN SOME WAYS the small intestine is actually larger than the long intestine. The small intestine, which is more than three yards long, is longer than the large intestine but narrower. Most nutrients are absorbed from digested food in the small intestine. The large intestine is shorter but wider. Its primary role is to form feces by removing water from undigested food.

STOMACH ACID is strong enough to dissolve a nail.

THE PANCREAS contains what are called the Islands of Langerhans.

THE AVERAGE American goes to the toilet an average of six times a day (for one reason or another).

SOME 75 percent of human feces is water.

Darwin-James theory fell out of favor among behavioral scientists and remained so, until recently.

Modern research into what is now called "facial feedback" began in 1984, when psychologists led by Dr. Paul Ekman from the University of California at San Francisco found that when people mimic different emotional expressions their bodies produce distinctive physiological patterns. These patterns include changes in heart and respiratory rate, changes in skin temperature and skin conductivity, and changes in general muscular tension level that differ from one emotion to the next.[15]

Dr. Ekman was the first to show through laboratory experiments that people actually feel happier when they smile, and feel sadder when their faces are "arranged" in what he calls a "configuration of sadness." Researchers at Clark University in Worcester, Massachusetts, led by psychologist Dr. James Laird, confirmed the finding that getting people to place the muscles of their face in the pattern of a given emotional expression actually produced that feeling. In several experiments, the expressions of subjects representing happiness, sadness, anger, and disgust all elicited the moods they portrayed.

In another study, University of Michigan researchers, led by psychologist Dr. Robert Zajonc, had subjects repeat vowel sounds over and over. When subjects pronounced either a long "e," which forces a mild smile, or an "ah," which imitates an expression of surprise, pleasant feelings were reported. But when subjects pronounced a long "u" or an umlauted German "u," both of which force a mild frown, the opposite emotions were reported—unpleasant feelings.[16]

Dr. Zajonc is a leading proponent of the Darwin-James theory. His studies also show that when facial muscles relax they raise the temperature of the blood flowing to the brain. But when the muscles tighten, they lower blood temperature. He theorizes that these temperature changes affect the activity of brain centers that regulate emotion. However, the real impact of these very subtle temperature changes remains very controversial. Dr. Zajonc says, "I'm not saying that all moods are due to changes in the muscles of the face, only that facial action leads to changes in mood."

Dr. Laird goes even further. He believes that facial expression not only alters mood but even alters memory. "If you're in a good mood," he says, "you remember positive things better than negative things. The effect is very specific when you're in an angry mood, you remember angry things but not necessarily sad things." He adds: "One way to manipulate mood is to manipulate facial expression. I had subjects read a Woody Allen story and an anger-provoking editorial. When I asked them to smile and recall what they'd read, they remembered more about the Woody Allen story. When I asked them to frown and recall what they'd read, they

THE LARGEST muscle in the human body is the gluteus maximus (the buttocks). It is also probably the strongest, based on the logic that girth is proportionate to strength.

THE LONGEST muscle is the sartorius muscle of the thigh.

APPROXIMATELY 1,500 surgical tools are left in the bodies of patients in the United States each year. Fatter patients are more prone to having a surgical tool left inside because of all the additional space in their bodies.

A PERSON WILL burn about 7 percent more calories walking on hard dirt than on pavement.

remembered the editorial better. Produce the face that matches the mood of the content and you'll remember more."

Precisely what physiological mechanisms are involved, and how important is this phenomenon to emotional life? The next research phase will address these questions and hopefully achieve some answers. Until then, Mona Lisa's secret is still safe.

There are still many mysteries in a smile.[17]

Why do I laugh when I breathe "laughing gas"?

"Laughing gas" is the common name for nitrous oxide, a gas discovered in 1772 by the British scientist Joseph Priestly (1733–1804). Priestly was also the co-discoverer of oxygen.

Nitrous oxide reduces the oxygen-carrying capacity of the blood, and this lack of oxygen (hypoxia) makes us feel giddy and "high." Inhaling nitrous oxide was found to produce a moderate euphoria, a light-headed giddiness that became a novelty at fashionable European parties of the nineteenth century.

Nitrous oxide is but one of many gases that can function as an anesthetic. It works by putting the nervous system more or less to sleep, so to speak, thereby more or less obliterating consciousness and thus "killing" pain. Although Humphry Davy (1778–1829) noticed that nitrous oxide blocked pain and therefore might be useful in surgery as an anesthetic, it was only in the 1840s that nitrous oxide was applied to medicine. Before then, all sorts of substances were used as painkillers—alcohol, opium, and mandrake, to name a few. Today, nitrous oxide is still used, particularly in dentistry.

Why can't I tickle myself?

Tickling is one of the least understood of all human physiological reactions. The "tickle response" is involuntary by definition. Although a person can sometimes control the tickle response by concentrating very hard, it cannot be self-induced.

It has been suggested that tickling with gentle movements of the fingertips excites certain small, fine nerve endings or touch sensors just beneath the surface of the skin. These are located all over the body, but especially on the palms and soles. The most obvious and observable reaction to tickling is laughter. However, the pulse quickens, blood pressure rises, and the body becomes keyed up and alert as well.

BLOOD IS THICKER than fresh water, but about the same thickness as sea water.

HUMAN ADULTS breathe about twenty-three thousand times a day.

MOST PEOPLE can carry a weight approximately equal to their own. By contrast, ants can carry a weight one hundred times heavier than themselves.

THE AVERAGE person will take in five pints of water a day. Three pints of that water come from drinking and two pints come from food.

IF YOU TOOK ALL the urine the world produces in one day, it would take a full twenty minutes to flow over Niagara Falls.

According to Dr. Roger Grief (an ironic name for an expert on tickling), professor emeritus of physiology and biophysics at Cornell University Medical College in New York, the fact that we cannot tickle ourselves is only one of a number of odd things about tickling. Another oddity is that the tickle response is ambivalent—that is, although the first reaction to tickling is usually pleasure, that pleasure sometimes becomes tinged with anxiety. Hence, the familiar expression "tickled to death" reflects something of the mingled fear and delight that tickling evokes.

According to Dr. William Fry, professor of clinical psychiatry at Stanford University Medical School, "if there is no anxiety or danger, people do not laugh or giggle when tickled. Nor will they laugh if they are tickled aggressively enough that they sense danger. People will laugh and giggle if they feel some anxiety but no danger."

If we try to tickle ourselves, we know that at any moment we can stop the stimulation, thus eliminating an essential component of tickling—our anxiety.[18, 19]

Why doesn't my belly button heal over?

This is one of *the* classic OBQs of all time. Isn't it easy to imagine this being asked by Neanderthals fifty thousand years ago? the ancient Greeks? medieval mystics? Himalayan monks? kindergarten children—and other great philosophers? The answer is probably not worth all the contemplation.

A belly button, or navel, is merely the umbilical cord scar tissue where the cord was detached following birth. Because it is of no medical significance, medical and anatomy texts pay little or no attention to it.

Nevertheless, any animal that has been nourished in the womb must have a belly button, although it may not always be easily seen. According to Dr. Edward Feldman, a professor of animal reproduction in the School of Veterinary Medicine at the University of California at Davis, "the scarring may be less obvious in some animals than in others, especially if it is covered by fur." The belly button does not heal over, because there is nothing between it and your stomach except a few thin layers of skin.[20, 21]

Why do I get motion sickness?

It's enough to make you throw up! Although many theories exist, medical experts admit they do not fully understand what causes motion sickness. Why some individuals seem to be more motion-sickness prone than others is particularly puzzling. Nausea, paleness, vomiting, sweating, and dizziness are all signs of motion sickness. About 90 percent of us suffer from motion sickness at least once in our lives.[22]

Dr. Mohamed Hamid, a vestibular disorders specialist at the Cleveland Clinic Foundation, claims that motion sickness probably originates

TEENAGERS catch colds twice as often as people over the age of fifty.

THE LIVER IS THE largest of the body's internal organs.

KIDNEYS ARE organs that clean the blood. Almost everyone has two kidneys, but most of us could survive with only one. In 1954, a patient of the Drs. J. Hartwell Harrison and Joseph Murray was very ill because he had no working kidney at all. He needed a new kidney urgently— one that was very like his own so his body would not reject it. Luckily for him, he had an identical twin. In the first truly successful transplant of its kind, his twin gave him one of his kidneys— and saved his life.

from one of two sources: problems in the inner ear (which controls balance) or problems in the central nervous system (which transmits electrical impulses to the brain concerning body and head movements). According to this theory, fluid "sloshes" through the ear's three semicircular canals simultaneously, resulting in contradictory nerve impulses reaching the brain, which in turn causes disequilibrium. Anxiety, stress, and fatigue are contributing factors.[23]

Dr. Cecil W. J. Hart, former chairman of the American Academy of Otorhinolaryngology, says: "Many people suffer from motion sickness because, when travelling, familiar cues used for orientation are upset." For example, "when a child is reading while riding in a car— the inner ears detect the motion of travelling, but the eyes see only the pages of print." Dr. Hart recommends that strong odors and greasy or spicy foods be avoided before or during travel.[24]

Dr. Kenneth Koch, a gastroenterologist at the Hershey Medical Center in Hershey, Pennsylvania, suggests that there may be a motion sickness gene that makes some people more prone to the sickness than others. This is based on analyses of the perceptual, neurological, and hormonal components of this loathsome gastric disturbance.[25]

Dr. Noel Cohen, an ear, nose, and throat specialist at New York University, adds that people subject to motion sickness should breathe fresh air and avoid alcohol and caffeine. For severe cases, he recommends medication taken orally or by transdermal patch.

Dr. Harold Silverman, a New York clinical pharmacologist and author of several health books, recommends the following:

☞ In an airplane, try to get a seat over the wing. There is less bouncing around from turbulence in that part of the aircraft.

☞ On a ship, try to stay on deck as close to the middle of the ship as possible, and don't focus on the motion of the waves.

☞ In a train, bus, or car, try to sit facing forward and focus your gaze straight ahead.

☞ Avoid reading during the trip.

☞ Avoid heavy meals and alcohol.

☞ If you take motion sickness medication, do so at least thirty minutes before you travel and every four to six hours during travel.[26]

Nevertheless, Dr. Kenneth Dardick, an inner-ear specialist at the University of Connecticut School of Public Health, warns that no medication has yet been developed to cure motion sickness. The best you can do is ride it out as "tolerances vary greatly from one person to another."[27]

Why don't people who take
nitroglycerin for heart conditions explode?

Evacuate! Evacuate! There are bombs bursting all over the cardiac ward!

We all know that nitroglycerin is a highly explosive compound. It is a volatile chemical cocktail combining carbon, hydrogen, nitrogen, and oxygen. "Nitro" taken in pill form helps heart patients by acting directly on the wall of the blood vessels. But according to Dr. Thomas Robertson, chief of the cardiac diseases branch of the National Heart, Lung, and Blood Institute of the U.S. National Institutes of Health, the amount and concentration of "nitro" in heart medications is too small to cause any possible danger of a patient exploding. This is so even if the patient overdosed a little and jumped up and down.

Dr. Robertson adds that "nitro" is "diluted with filler in the tablet," such that by the time "it is absorbed in the body it is in minute concentrations. It dilates the vessels, which both increases the blood supply to the heart and reduces the work of the heart by reducing blood pressure."[28]

TO MAKE ROOM for the heart, the left lung is smaller than the right lung.

OUR HEARTS pump about forty million gallons of blood in a lifetime. Your heart could fill a swimming pool in about twenty-five days if you had enough blood to spare. At any one time, your body contains just a little more than one gallon of blood.

What causes different
blood groups?

Most of us know that there are four major human blood types: A, B, AB, and O. Each is divided into Rh+ and Rh–, which makes a total of eight major blood groups in all. This is called blood polymorphism.

Despite research, science does not know why we have only eight and not eighty, eight hundred, eight thousand, or even only one. For survival as a species, it seems to make little sense to have more than one blood group. That is because there appear to be obvious disadvantages but no advantages. For example, the best-known disadvantage arises when Rh– mothers have children by Rh+ fathers. The mother-baby incompatibility may lead to an infant's death. It seems that evolutionary pressure should have long since eliminated this incompatibility problem, but that did not happen—and it baffles science.

Disadvantages are often compensated for by advantages. For example, the hemoglobin polymorphism that gives rise to sickle-cell anemia in Central Africa populations also helps protect those populations from malaria. The trade-off is referred to as "balanced polymorphism."

Unfortunately, science has not been able to establish that all blood polymorphisms are balanced. In fact, we know little about nature's blood-balancing act, but we can observe that some bloods have some disadvantages over others. For example, according to research by Dr. Corrine Wood of the Department of Family Medicine at the University of California at Irvine, people with Type O blood, compared

IT TAKES SIXTY seconds for blood to make one complete circuit of the human body.

THE HUMAN body makes about two million new red blood cells every second.

THE DAILY HEAT output of the average person would boil eight gallons of freezing water.

YOU WOULD need to walk thirty-four miles to melt away one pound of fat.

IF YOU COULD stretch out all of a human's blood vessels, there would be more than enough to go around the world twice.

with those who have A or B or AB blood, are more prone to typhoid fever, virus diseases (especially polio), bleeding, autoimmune diseases, and gastric ulcers. Also, malaria-carrying mosquitoes prefer to take blood from Type O individuals, although why remains unclear.[29]

According to research by Dr. G. Jorgensen of the Frederick Cancer Research and Development Center of the National Institutes of Health in Bethesda, Maryland, people with Type A blood, compared with those who have O, B, or AB, are more prone to malaria, cancer, smallpox, diabetes mellitus, cardiac infarction, pernicious anemia, rheumatic diseases, and nephrolithiasis (a kidney problem involving the buildup of salts).[30]

But we can go only so far with such comparisons. Drs. J. A. Beardmore and F. Karimi-Booshehri, of the Department of Medical Biochemistry and Genetics at the Tehran University of Medical Sciences in Iran, reported that Type A blood was significantly more common among those in higher socioeconomic groups in the United Kingdom, compared with those who have Type O blood. It is interesting that a torrent of criticism followed this report.[31]

The existence of blood polymorphism is a fact. Its meaning and biological significance remain mysteries.[32]

Can a person have more than one blood type?

This raises the subject of the fascinating phenomenon of the blood chimera.

We were all taught in high school biology that each of us has one blood type—A, B, AB, or O—and *only* one type. Moreover, we were taught that it is impossible for a person to be of two blood types and that the body would reject any incompatible blood type taken in transfusion. But this does not account for blood chimeras.

A blood chimera is a human with two different blood types plus the tissues for manufacturing new blood cells for both types. Blood chimeras occur in animals and in humans, although very rarely. All known human blood chimeras are twins. It is believed that somehow, while still in the womb, three very odd things happen: Blood is shared between the twin fetuses, blood-manufacturing tissues are exchanged between them, and the immune system's normal rejection response to foreign blood is suppressed.

If your blood is Type A, your blood carries a protein called antigen A and another protein called antibody B. An antigen is a substance that stimulates the body to produce an antibody. If your blood is Type B, your blood carries antigen B and antibody A. If your blood is Type AB, your blood carries both antigens but neither of the antibodies. Finally, if your blood is Type O, your blood carries neither of the antigens but both of the antibodies.

FRENCHMAN Michel Lotito has a very unusual diet. Born on 15 June 1950, he has been consuming large quantities of metal and glass since he was nine years old. To date, he has eaten supermarket carts, television sets, bicycles, chandeliers, razor blades, bullets, nuts and bolts, lengths of chain, computers, and an entire Cessna 150 light aircraft, which took him nearly two years to consume.

Because of this, people with Type A blood can safely receive blood from Type A or Type O people. People with Type B can safely receive from Type B or Type O. People with Type AB (the universal recipient) can receive blood from any of the other three because their blood is compatible with all the others. People with Type O (the universal donor) can give blood to all the others but ironically can receive blood only from other Type O's.[33]

chapter 10
endings

Death happens to us all. But there's more to death than simply, as Shakespeare wrote, "shuffling off this mortal coil."

Can people die of fright?

A legal opinion does not make a physiological fact, but an increasing number of lawyers and jurists now believe that fear is entirely capable of killing. Dr. George Engel, professor of internal medicine at the University of Rochester, claims: "These lethal reactions are rare, but they do happen. Most people don't drop dead in the face of a shock. Some do."

More people are dying of fright because there has been an increase in such frightening crimes as robberies and assaults, and because in general people have more things to be afraid of. In a common

HYPOXEMIA—
lack of blood to
the brain—is the
cause of every
human death.

MEN ARE SIX
times more likely
than women to
be struck by
lightning.

scenario, an elderly person is frightened by an attempted robbery and dies of heart failure. Damage to the heart can occur quickly during periods of overexertion. It can also be triggered by one's state of mind.[1]

Can you die of laughter?

For the same reason, you can die of laughing. The Australian Associated Press reported one case in which a Danish doctor literally died laughing when a fit of laughter brought on a fatal heart attack at a cinema. The doctor was watching *A Fish Called Wanda* (1988). It was believed that his heartbeat accelerated from a normal rate of 60 beats a minute to between 250 and 500 beats.[2]

Does your hair continue to grow after you die?

The conventional wisdom is that hair only *seems* to grow. As with supposed fingernail and toenail growth after death, hair looks as if it has continued to grow because after death the body dries and shrivels up. The skin contracts and more of the hair is revealed.

However, there are rare cases of so-called hair "revelation" after death far beyond that which can be attributed to mere skin contraction. For example, in their classic work *Anomalies and Curiosities of Medicine* (1897), George Gould and Walter Pyle wrote: "Aristotle discusses postmortem growth of the hair, and Garmanus cites an instance in which the beard and hair were cut several times from the cadaver. We occasionally see evidence of this in the dissecting rooms. Caldwell mentions a body buried four years, the hair from which protruded at the points where the joints of the coffin had given away. The hair of the head measured eighteen inches, that of the beard eight inches, and that of the breast from four to six inches."[3]

There are other such cases, most of which are a century old or more. How much credibility we can give to such accounts is up for debate.

THE AVERAGE age at which one is likely to be struck by lightning is thirty-five.

AMERICANS choke to death on toothpicks more than on any other object.

Does hair color change after death?

Hair color can change in response to the immediate chemical environment. That is why hair dyeing is so easy and so common. Sometimes hair dyeing is unintentional. For example, workers in cobalt mines may find their hair has turned blue, and workers in copper mines may have their hair turn green.

But hair dyeing after dying is a horse of a different color. It seems that it is possible, though not very common. A case originally reported in *Popular Science Monthly* in 1885 is one possible instance of hair color changing after death. A doctor by the name of Hauptmann described a body exhumed after being buried for twenty years. The hair had changed from dark brown to red. Brown is close to red on the color spectrum, so whether the change was merely the fading of brown to red is worth considering.[4]

Can humans "glow in the dark" after death?

A dead body can sometimes look as if it is glowing when it really isn't. One example of this observation comes from the book *Phenomena: A Book of Wonders* (1977) by John Mitchell and Robert Rickard: "As neighbors prepared the shroud, they noticed the body surrounded by a blue glow and radiating heat. The body appeared to be on fire; efforts to extinguish the luminescence failed, but eventually it faded away. On their moving the body, the sheet below it was found to be scorched."[5]

It all looks intriguing, but the explanation is rather simple. After death, the human body decomposes. Bacteria play a major part in this natural process, and some bacteria are luminous. Thus, in the rare case that a decomposing human body appears to be glowing, it is probably merely bacteria.

That explanation was first presented in 1838 by Daniel and Robert Cooper, who conducted experiments to test various theories about why dead humans might seem to glow like glow-worms. It seems they were very curious about the "fuel" as well as the "fire."[6]

Of course, if the person died by swallowing radium, they really would glow in the tomb.

Do I have a dead twin that I
don't know about?

This is a real possibility. Recent advances in medical technology make it possible for us to know that far more twins are conceived than are actually born: One dies in utero. This is known as the "vanishing twin phenomenon."

What happens to the never-born twin? How often does a twin vanish?

Although first described by researchers in the 1970s, the "vanishing twin" phenomenon has been little understood until recently. Its existence is now firmly documented by ultrasound scan analysis.

Based on the same principles used in sonar during World War II to detect enemy submarines, ultrasound scanning yields a visual image on a television-like monitor showing the developing fetus from conception to birth. In fact, this image is so clear that fetal growth can be precisely measured, fingers and toes counted, fetal position shifts noted, and even urine observed in the fetal bladder. Invaluable as a medical diagnostic tool, ultrasound is like watching your unborn baby on television (black and white only, of course, and with the volume turned off!). Many doctors now have portable ultrasound units in their offices.

Almost from the time ultrasound was first used, doctors began noticing that not all twins that were conceived were born. Ultrasounds early in pregnancy showed twins, yet later ultrasounds revealed only one fetus. Something was happening, but what? Further observations established that, whatever was happening, the disappearing twin just vanished sometime during the first month of pregnancy—and never after the first trimester.

About a dozen "vanishing twin" studies appeared in the medical literature beginning in 1979. The vanishing was attributed to natural absorption or rejection processes of the mother's body. However, what was more intriguing about the studies was the estimations of twin disappearance rates. While some researchers denied the existence of the phenomenon, others estimated the rate of twin conceptions to be as high as 78 percent.

Several factors contributed to the wide range in the estimations, including differences in study technique, methodology, and sample size, and incorrect interpretations of technical ultrasound features or of physiological conditions associated with pregnancy, which can actually mimic the presence of an additional gestational sac and thus fool the ultrasound observers.

This was further complicated by a number of confounding variables. For example, rates of twin births vary among nations. The reported incidence of twin births is highest among blacks and East Indians, followed by Northern European whites. Mongolians have the lowest rate. In the United States, twins are born about once in eighty-five births. In Australia, twins are born about once in every one hundred births. Some other national data are: 1 in 80 in the United Kingdom, 1 in 86 in Italy, 1 in 130 in Greece, 1 in 150 in Japan, and 1 in 300 in China.

In one U.S. study, where one thousand pregnancies were closely monitored via ultrasound and carefully analyzed, there was a "minimum incidence" of conceptual twinning, between 3.29 percent and 5.39 percent—"higher than previously believed." Furthermore, it was found that vanishing occurred in 21.2 percent of conceived twin cases—again higher than previously believed.[7]

Vanishing is often accompanied by vaginal bleeding. The theory is that the vanishing occurs when the mother's body either absorbs or rejects a fetus due to severe genetic or other defects in the fetus. However, the remaining twin is not genetically or otherwise defective, is left perfectly intact, and has every chance of being born completely healthy. The "half miscarriage," so to speak, explains the vaginal bleeding.

The "vanishing twin" process is nature's way of protecting an individual who is capable of surviving even if its twin is not. The alternatives are harsher.[8]

Can you have "an attitude" after death?

"**A**ttitude" in this context refers to the position in which a corpse can be frozen after sudden death. Sometimes the body assumes a truly bizarre attitude at the instant of death and remains unchanged, with no evidence of muscle relaxation. This phenomenon is often observed on battlefields. It is quite different from normal rigor mortis.

When a body dies, it is usually quite flexible and limp. Then rigor mortis sets in, and the muscles of the body gradually stiffen, then gradually relax again. The cycle of rigor mortis may take twelve to twenty-four hours or more. The length

A SINGLE ORGAN transplant donor can provide organs, bone, and tissue for fifty or more people.

THE OLDEST authenticated pair of female twins were Kim Narita and Gin Kanie of Japan. They were born on 1 August 1892. Kim died of heart failure in January 2000 at 107 years of age.

of time depends on certain conditions, such as the cause of death, the air temperature, and so on.

Nevertheless, in some cases of sudden and violent death a body will freeze in the position the person was in at the very instant of death, and for some unknown reason the muscles fail to relax, remaining tense indefinitely.

LEPROSY IS THE world's oldest disease, with documented cases dating back to 1350 B.C. Leprosy is now 100 percent curable.

THE LONGEST-living triplets were Faith, Hope, and Charity, who were born in Elm Mott, Texas, on 18 May 1899. Faith was the first to die at age ninety-five in October 1994.

This phenomenon of "strange attitude at death" was frequently observed in nineteenth-century journalistic accounts of wars. The trench warfare of World War I provided so many cases that such gruesome descriptions were simply no longer thought worth telling.

Perhaps the best description of such an unusual attitude in death on the battlefield is the following: "[It] was observed by Dr. Rossbach, of Würzburg, upon the battlefield of Beaumont, near Sedan, in 1870. He found the corpse of a soldier half-sitting, half-reclining, upon the ground, and delicately holding a tin cup between his thumb and forefinger, and directing it toward a mouth that was wanting. The poor man had, while in this position, been killed by a cannonball that took off his head and all of his face except the lower jaw. The body and arms at the instant of death had suddenly taken on a rigidity that caused them to afterward remain in the position that they were in when the head was removed. Twenty-four hours had elapsed since the battle, when Dr. Rossbach found the body in this state."[9]

Is there such a thing as a "near-death experience"?

There's a great deal of confusion about the so-called "near-death experience" (NDE). Medical and behavioral scientific research is divided as to whether whatever is experienced is a hallucination, a product of mass hysteria, a chemical reaction in the brain due to traumatic stress, or truly a stage between life and death.

Improved medical technology and treatment have brought more people back from death's door than ever before. As a result, more survivors are reporting near-death experiences. Usually they are patients who survive cardiac arrest, coma, near fatal trauma, near drowning, or some other severe illness. Yet there are also cases of NDEs resulting from catastrophic psychological stress not involving physical injury. Examples of this are survivors of mountain-climbing accidents and miners trapped for days after cave-ins.

Probably the first case of a near-death experience was reported by the ancient Greek philosopher Plato (427–347 B.C.). In *The Republic,* Plato tells of a soldier named Er who was supposedly killed in battle. As the account goes, while Er lay on the battlefield, his soul took flight. Accompanied by the souls of several fellow soldiers, Er's soul experienced another world—a land where all souls were judged. He saw other souls choosing their subsequent incarnations and then drinking from the River of Forgetfulness in order to obliterate their past memories. But Er was forbidden to drink, and then he blacked out. He revived into his real, conscious life just in time—as his funeral pyre was being lit.[10]

In 1975, Dr. Raymond Moody wrote a book titled *Life After Life* in which he coined the phrase "near-death experience." This term was used to describe "the common pattern of elements that people report experiencing after they have recovered" from a close brush with death. Dr. Moody's book was filled with case after case of NDEs.[11]

It is now generally accepted that a near-death experience includes any memory a person has from the time they were unconscious—lingering between life and death—and the time they regain consciousness.

In a near-death experience, one or more of nine elements are experienced:

1. *A sense of being dead.* The sudden awareness that one has had a "fatal" accident or not survived an operation.

2. *Peace and painlessness.* A feeling that the ties that bind one to the world have been cut.

3. *An out-of-body experience.* The sensation of peering down on one's own body and perhaps seeing doctors and nurses attempting resuscitation.

4. *Tunnel experience.* The sense of moving up or through a narrow passageway.

5. *"People of light."* Being met at the end of the tunnel by others who are "glowing."

6. *A "Being of light."* The presence of a God-like figure or force of some kind.

7. *Panoramic life review.* Being shown one's life by the "Being of light."

8. *Reluctance to return.* The feeling of being comfortable and surrounded by the "light," often described as "pure love."

9. *Personality transformation.* A psychological change involving loss of the fear of death, greater spiritualism, a sense of "connectedness" with the earth, and a greater zest for life.[12]

Of these, the most frequent element in the near-death experience is by far the out-of-body experience. Perhaps 75 percent of reported NDEs involve an out-of-body experience. The second most frequent element is the so-called "panoramic life review." As the term suggests, the events in one's entire life flash by in the mind in an instant. In order of diminishing frequency, the other common features of NDEs include entering a tunnel, meeting others (such as living or dead relatives), encountering "a Being of light," having a sense of the presence of "a deity," returning to the body, and experiencing what researchers call "elements of depersonalization" (such as an altered sense of time or a detachment from reality).

Another interesting finding is that there is an extraordinary similarity in the backgrounds of those who report having a near-death experience: There is a "consistent tendency" for these people to have also reported being victims of child abuse.[13, 14]

Many theories explaining near-death experiences abound, and many studies are teasing out the myths from the reality. Another interesting finding in recent years is that there is a large cultural component in NDEs. Which elements are more likely to appear differs according to the society in which a person was brought up. For example, based on the work of Dr. Nsama

THE WORLD'S oldest quadruplets were the Ottman siblings: Adolf, Anne-Marie, Emma, and Elisabeth. They were born on 5 May 1912. Adolf died first at age seventy-nine in March 1992.

Mumbwe of the Department of Psychology at the University of Zambia, it has been noted that many Africans reporting an NDE see the experience as evil and involving witchcraft. Among Japanese who have an NDE, many report seeing long, dark rivers and beautiful flowers, which are common symbols in Japanese art. And East Indians sometimes see a "bureaucracy" in their vision of heaven. Finally, Micronesians sometimes see heaven as a large city with skyscrapers.

Dr. Kenneth Ring has identified twelve changes that often occur in a person after a near-death experience:

1. A greater appreciation of life
2. A higher self-esteem
3. A great compassion for others
4. A heightened sense of purpose and self-understanding
5. A desire to learn
6. An elevated spirituality
7. A greater ecological sensitivity
8. A feeling of being more intuitive, sometimes "psychic"
9. An increased physical sensitivity
10. A diminished tolerance to light, alcohol, and drugs
11. A feeling that his or her brain has been "altered" to encompass more
12. A feeling that he or she is now using the "whole brain" rather than just a small part[15, 16]

Is my chance of dying related to my astrological sign?

Astrology is pseudoscience, but tell that to the millions of people who follow their star sign in newspapers every day! Or tell it to the newspaper editor who neglects to include the astrology column!

Is it possible to be born under a bad astrological sign? Do people born under one sign live longer than people born under some other sign? Generally speaking, there is no such relationship. However, believe it or not, two behavioral scientists have uncovered evidence that an astrological sign can have a bearing on your death in one way. They have shown that one's astrological sign can predict a person's inclination toward considering suicide ("suicide ideation").

Specifically, it's the sign of Pisces that seems to be the naughty one. Statistically, according to the two researchers, Pisces is "significantly associated with suicide ideation."

Dr. Steven Stack from the Department of Sociology at Auburn University in Alabama, and Dr. David Lester from the Department of Psychology at Stockton State College in New Jersey, published their findings in an article under the intriguing title of "Born Under a Bad Sign? Astrological Sign and Suicide Ideation." Although their article generated little initial attention, the implications of their findings refuse to go away.

Traditional scientists—those who unswervingly practice the scientific method, demand hard empirical evidence at all times and scoff at the mere suggestion that astrology could have any possible predictive value for humans—have been forced to confront the findings of Drs. Stack and Lester and perhaps even reconsider long-held

THE MAN WITH the oldest known individual ancestor is British Professor Adrian Targett. DNA tests matched Professor Targett's with that of a 9,000-year-old skeleton found in Cheddar, England.

CAN LIGHTNING strike a human twice? Yes, it can. In fact, one man, Roy Sullivan, a park ranger from Virginia, was known to be struck not only twice, but seven times. The strikes occurred between 1942 and 1977. Ironically, having survived these strikes, this human lightning rod took his own life in 1983, reportedly over a failed romance.

positions. After all, part of being a scientist is keeping an open mind—being open to new evidence.

"Suicide ideation" refers to the tendency to think positively about suicide and to seriously consider attempting suicide —regardless of whether the act is actually carried out. Suicide ideation is often the first indication of a subsequent suicide attempt. As such, it has been the object of much behavioral science attention.

Previous international research on suicide ideation has revealed some interesting findings. For example, approval of suicidal behavior is more pronounced among Catholics than Protestants, among young people more than the middle-aged or the elderly, and among people who, when young, had negative relationships with their parents.

Until the Stack-Lester study, astrology was never investigated to see whether there were any links with suicide. Applying sophisticated statistical techniques to 7,508 subjects drawn from the General Social Survey of the Roper Public Opinion Research Center in Storrs, Connecticut, the two researchers found that "while astrological sun sign on the whole was not predictive of suicide attitudes, . . . the sun sign of Pisces is associated with suicide attitudes." On four indicators of suicide ideation, "respondents with the Pisces sign were more approving of suicide than the rest of the population."

Drs. Stack and Lester suggest a theory to explain their findings. In both the Greek and the Indian interpretations of astrology, people born under the Pisces sign "will have lives characterized by loss and may, indeed, feel more depressed and hopeless. If such were the case, it would be anticipated that persons in such a depressed mood would be more approving of suicide." The two researchers conclude: "Whether this reflects astrological phenomena *per se* or socialization factors wherein the reality of a bad sign becomes a self-fulfilling prophecy is beyond the scope of the present analysis."[17]

Another possible interpretation might be that the findings of Drs. Stack and Lester on Pisces is merely a statistical quirk, an anomaly, a chance, random finding without causal significance or meaning. Such quirks are common in the mathematics of behavioral science research.

It has been said that if you have enough chimpanzees and enough paints, brushes, canvases, and enough time, one of the chimpanzees will eventually paint the Mona Lisa.

But don't hold your breath.[18]

Can a human being burst into flames for no apparent reason?

This refers to "spontaneous human combustion" (SHC), which applies to unexplained deaths or injuries by fire where the source of the fire is apparently unknown. SHC has become a favorite subject of many devotees of paranormal phenomena.

In fact, science shows us there is no such thing as spontaneous human combustion and that on further investigation all such supposed cases have clear explanations.

The first supposed case of spontaneous human combustion in the medical literature was reported by a Dr. John Overton in an 1836 article in *Transactions of the Medical Society of Tennessee*. The article is about burn injuries of seemingly unknown origin suffered by a mathematics professor in 1835.

In *Bleak House* (1852) by Charles Dickens, the character Krook meets a macabre death by spontaneous human combustion. Dickens was roundly criticized in his day for (dare we say) fanning the flames of public fears concerning this phenomenon, which was wholly unproven then as now.

Periodically, a case of alleged spontaneous human combustion is reported in the medical literature and more often in the public press. For example, a 1990 report from Beijing began: "Chinese doctors are alarmed at a new medical mystery—a young boy whose body can ignite spontaneously in the most sensitive places, burning through clothing." The report added that the boy's armpits, right hand, and "private parts" were continually being burned.[19]

Perhaps the most famous case of this sort occurred in England in 1985. It is described at some length in *Death by Supernatural Causes* by Jenny Randles and Peter Hough.[20] In this case, after attending a cooking class a teenage girl burst into flames while descending a stairway. Although she received only superficial burns to her back, she died in hospital two weeks later of complications related to infection.

Far from being an unexplained death by spontaneous human combustion, the probable explanation is far more mundane. Dr. Bernard Knight, a professor of forensic pathology at the University of Wales College of Medicine, writes in *New Scientist* about this case and dismisses all cases of alleged SHC.[21] According to Dr.

Knight, "the interest is not in that tragedy, which seems a clear instance of a gas jet igniting her clothing—but in the limitless need for people to believe in spontaneous human combustion, a fantasy stimulated by Charles Dickens in *Bleak House*. Denied by pathologists for years, spontaneous combustion is a mass delusion that refuses to go away."

Dr. Knight adds: "Certainly, dead bodies can catch fire and burn almost to nothing—but there must be a good source of ignition and a good draught, so that the body fat acts as fuel for the wick of clothing. There is nothing 'spontaneous' about this, and the girl whose clothing caught fire at the back must have been too near the gas rings, albeit some minutes before, leading to minimal smoldering before flames developed."

Hardly a paranormal phenomenon, the notion of spontaneous human combustion is one that science shoots down in flames.[22]

Is my chance of dying related to the phase of the moon?

The power of the full moon has been blamed for bringing about insanity, accidents, suicide, and murder, but is there a scientific foundation for the power of the full moon to significantly affect the way we behave?

STRICTLY speaking, nobody is "buried" in Grant's Tomb. President Ulysses S. Grant and Mrs. Grant are "entombed" in New York City. A body is "buried" only when it is placed in the ground and covered with dirt.

THE ODDS OF being killed by falling space debris are one in five billion.

The belief that humans are somehow affected by the full moon is one of our civilization's oldest notions about the causes of behavior. Both Pliny the Elder (A.D 23–79) and Plutarch (A.D 46–120) wrote about the widespread nature of this belief. Anthropologists report that it has been an extremely common notion throughout the belief systems of many non-Western societies. Our traditional folklore is filled with such references, as is our popular culture today.

Staff at psychiatric hospitals around the world have for many years claimed that patients are usually more difficult to manage during nights with a full moon. The term "lunatic" is sustained in our language by the persistence of such reports.

THE ODDS OF being struck and killed by a meteorite are about one in ten trillion.

THE ODDS OF being killed by a dog are about 1 in 700,000.

IT HAS BEEN calculated that in the last 3,500 years there have been only 230 years of peace in the civilized world.

We know that the full moon affects tides, plant growth, and many other physical and biological processes. Women's menstruation, possibly hair and nail growth, and perhaps other aspects of human biology may be influenced as well. Statistically, more babies are born during the full moon than at any other time in the lunar cycle. Yet the reasons for this remain unclear.

Research on the possible effect of the full moon on erratic or violent behavior has been inconclusive over recent decades. Possibly the most convincing evidence that lunar cycles affect violent behavior comes from Dr. Arnold Lieber of the University of Miami School of Medicine. In one study, Dr. Lieber concluded that the occurrence of self-destructive acts is positively correlated with the full moon. It was then theorized that this must be due to some inherent "biological rhythm of human aggression."[23]

Other studies hint at possible relationships but fail to establish statistically significant correlations.

Still other studies have completely rejected the full moon as an influence on erratic or violent behavior. And this has been the trend in the research for many years. One study has statistically demonstrated that the full moon did not affect the U.S. homicide or suicide rates.[24] In another study, it was shown that the phases of the moon did not alter the rate of disruptive behaviors of inmates at a U.S. psychiatric institution.[25]

In an unusual twist, still another study found that the number of patients with violent injuries actually fell during the full moon. This goes 180 degrees against the commonly held view. In this study, Drs. Wendy Coates, Dietrich Jehle, and Eric Cottington from the Allegheny General Hospital in Pittsburgh reviewed all admissions for trauma over a period of one year at their hospital. They found that of the 199 stabbing and shooting victims admitted, there were eight patients admitted on full-moon days for every ten admitted on days when there was no full moon. This led them to conclude: "The belief in the deleterious effects of the full moon on major trauma is statistically unfounded."[26, 27]

Is my chance of dying related to the activities of the sun?

Science knows of no direct cause-and-effect relationship between human mortality and the activities of the sun. Early attempts to find such correlations have been nullified by later research.

In 1972, two Russian scientists alleged that they had evidence linking solar activity with the behavior of various living organisms. However, this evidence was not made available outside what was then the U.S.S.R. They even proposed a new branch of science to study the solar-behavior relationship.

But that idea was quickly hit squarely on the head only a few years later. In 1976, Drs. B. J. Lipa and P. A. Sturrock of the Institute for Plasma Research at Stanford University, along with Dr. F. Rogot of the National Heart and Lung Institute at the U.S. National Institutes of Health, published impressive results showing that solar activities have no bearing on human mortality. They examined U.S. death rates and found no link with the sun whatsoever.[28]

Why do people "die in their sleep"?

There should be no real mystery to this. If people spend roughly eight hours a day sleeping, and if they die of "natural causes," then they have a one in three chance of dying during sleeping. But beyond this, there exists a strange mystery regarding sleep-related death that baffles medical science and defies explanation. It is called "Sudden and Unexplained Death in Sleep Syndrome" (SUDS).

SUDS occurs among adults, particularly among Asian adult males. No one knows why it is more common among men, or why Asians are the most susceptible. The U.S. Centers for Disease Control and Prevention in Atlanta labeled SUDS as the leading cause of death among young Southeast Asian men who came to the United States as refugees in the early 1980s. SUDS has even been referred to as the SIDS of adults. (SIDS refers to Sudden Infant

Death Syndrome, the leading cause of death in both the United States and Australia from birth to age one.)

The first report of SUDS in the medical literature appeared in 1917 in the Philippines, where it was known as *bangungut*. A 1959 report from Japan referred to the syndrome as *pokkuri*. It has been reported in Laos, Vietnam, Singapore, and elsewhere, and it has been known by various names. But it is the same strange, unexplained phenomenon.

The victims of SUDS are apparently in good health just before their deaths, so their tragic, sudden demise comes as a great shock to everyone. The family is often left destitute because it is often the male breadwinner who dies so suddenly.

The accounts of witnesses claim that the victims are apparently sleeping normally when all of a sudden they begin to moan, groan, snore strangely, gasp, and finally choke. Technically, these are termed "agonal signs." Most SUDS victims actually die of "ventricular fibrillation," sometimes within minutes of when the "agonal signs" first appeared. "Ventricular" refers to the small, lower chambers of the heart, while "fibrillation" is a local, involuntary contraction of a muscle, invisible under the skin.

Some witnesses have tried to awaken the obviously distressed victim-to-be, but that is invariably unsuccessful, and the person dies just the same. When an autopsy is performed in a SUDS case, it fails

IF WE HAD THE same mortality rate now as in 1900, more than half the people in the world today would not be alive.

FAMOUS LAST words: "It hurts." —French President Charles de Gaulle

CURRENTLY, only 1 person in two billion lives to be 116 or older.

to reveal any significant pathological conditions or evidence of accidental poisoning, allergy, or foul play.

So what *do* we know about SUDS?

Dr. P. Tatsanavivat and six colleagues from the Department of Medicine at Khon Kaen University in Thailand described their survey of SUDS deaths in northeastern Thailand over a two-year period. The typical pattern of SUDS in their survey is as follows: The victim dies within twenty-four hours of the onset of the "agonal signs," is usually between twenty and forty-nine years of age, has "no history of severe illness, was in good health during the previous year, and was able to work during the twenty-four-hour period before death."[29]

The researchers add: "The deaths were witnessed in 63 percent of cases and the other victims were found dead in a sleeping or resting position. In witnessed cases, 94 percent of deaths occurred within sixty minutes of onset of acute signs or symptoms. All of the reported SUDS victims were men." In addition, the victims were of normal body weight. Rates of smoking, drug usage, alcohol consumption, and other possible risk factors were also normal.

It is interesting that a family history of SUDS was reported in about 40 percent of cases. In fact, about 18 percent of victims had brothers who had also died suddenly—but none had sisters who had died in the same manner.

SUDS seems to be seasonal. In Thailand at least, it is most common during March through May and least common during September, October, and November.

The researchers note that SUDS is becoming recognized "as a potentially important public health problem" in Thailand. It kills about one in three thousand men between the ages of twenty and forty-nine and ranks as a killer in this group only behind accidents, poisonings, violence, and heart disease.

At the village level, superstition takes over in the absence of a reasonable explanation for SUDS. The researchers note that the people of rural northeastern Thailand refer to SUDS as *laitai* ("death in sleep"). The local explanation for *laitai* is that a "widow ghost" goes looking for the spirits of young men, waits for the young man to fall asleep, and then steals his spirit—causing sudden death.

The researchers note: "Fear of *laitai* and the 'widow ghost' is now widespread in northeastern Thailand and rituals have emerged that involve disguising sleeping men with women's cosmetics, fingernail paint, and bed clothes."

One scientific explanation for SUDS is that a combination of physical and psychological stressors may somehow be causing SUDS. For example, psychological factors causing related heart problems was put forward in one study, but others regard that view as highly speculative.[30]

"Widow ghost" or whatever, SUDS remains a mystery.[31]

What is a zombie?

Zombies, the so-called "walking dead," are an integral part of life in the Caribbean nation of Haiti. Among voodoo believers, zombies are reputedly corpses that have been

ANCIENT Egyptians used the herb thyme to help preserve mummies.

brought back to life through "black magic." The corpse is allegedly revived after the power of the snake or python deity enters the body. The corpse, now a zombie, is still regarded as a human being, but one without speech, will, or judgment. Nevertheless, the zombie is capable of automatic movement, such as walking and following simple commands. It is a trancelike, sleeplike state. Indeed, the term "walking dead" is very descriptive.

The origin of the belief in zombies is lost in antiquity, but it is almost certainly derived from the traditional religious and folklore traditions of West Africa, where Haitians came to the New World as slaves.

Although educated Haitians and Westerners regard zombies as imaginary beings, perhaps on a par with the werewolves of Romania or the leprechauns of Ireland, most Haitians regard zombies as very real. In fact, even today few Haitians dare to violate the voodoo taboo that forbids even mentioning the subject. Indeed, the making of zombies—"zombification"—is a term recognized in Haitian law.

Dr. Lamarque Douyon, a Haitian psychiatrist in Port-au-Prince, has studied the phenomenon of zombies for nearly three decades. Where other Haitian physicians refuse to attend zombies who also happen to need medical treatment (no doubt in fear of the voodoo taboo against doing this), Dr. Douyon has treated many such people. Placing science, understanding, and

ACCORDING TO British law passed in 1845, attempting to commit suicide was an offense punishable by death.

THE LONGEST coma from which a person has emerged lasted thirty-seven years.

devotion to humanity ahead of fear and superstition, he believes he has solved the mysteries surrounding zombies.

Dr. Douyon contends that making a zombie involves simple biochemistry and native folk pharmacology. He claims that turning a normal person into a zombie is merely a matter of drugging the person without his or her knowledge. In fact, zombification is generally a punishment imposed by the voodoo courts in Haiti. Haitians, it seems, as often as not turn to voodoo courts for justice in resolving a variety of disputes ranging from property matters to personal grievances, because the official Haitian legal system is seen as corrupt and serving only the rich. Thus, the voodoo court might punish a person by turning

WHEN IT comes to cannibals, "indocannibals" eat only people they know (that is, people inside their own group). "Exocannibals" eat only people they don't know (that is, people outside their group).

him into a zombie, in much the same way that an American court might impose a fine or period of imprisonment.

According to Dr. Douyon, transforming a person into a zombie is accomplished by having them drink one of several kinds of poison. The first, tetrodotoxin, has the effect of putting normal human beings in a deathlike trance. This poison is extracted from a ball-shaped fish, the tetrodon, and combined with severed parts of a dead toad for good measure. The second poison is made from a normally inedible vegetable appropriately named the "zombie cucumber" (*concombre zombi*). This poison has the effect of slowing the metabolism to the point where the victim appears to be dead. Yet another drug, still unknown to Dr. Douyon but well known to

the **odd body**

······················

YOU MAY NOT think of the flu as a big killer, but in 1918 to 1919 an outbreak of an unusually nasty flu virus killed twenty-five million people throughout the world. There were eighteen thousand deaths in London alone. The flu killed more people than World War I, which had just ended.

THERE ARE ABOUT twenty-five thousand people in the world who are at least one hundred years old.

voodoo priests, gives the already comatose-like person a more normal metabolism and physical functioning ability.

When the poison is administered, Dr. Douyon stresses, the victim soon appears to be dead. The bereaved family, unaware that the victim is still alive, often proceeds with burial. Later, after the zombie is revived, he or she returns in a disoriented, stuporous condition that gives rise to the myths of zombies "walking dead,"

Haitian funeral and burial traditions permit zombification to persist. In fact, proper etiquette calls for family and friends to leave the cemetery before the actual burial takes place. According to Dr. Douyon, this allows all manner of foul play to occur. A body may be switched, abducted, revived, and so on, without discovery.[32]

How do you make a dead body into a mummy?

Embalming and creating mummies originated with the Egyptians about 3000 B.C. "Mummification" comes from a Persian word meaning "was." The techniques of mummification were developed by trial and error. Earlier mummies tend not to be mummified as well as later ones.

The oldest known mummy is that of a young princess who was mummified around 2600 B.C. on a plateau near the Great Pyramid of Cheops at Giza. The mummy was excavated in 1989. The oldest complete mummy is that of a court musician who was mummified around 2400 B.C. in the Nefer tomb of Saqqara. It was unearthed in 1944.

Generally speaking, it is believed that the corpse was preserved with aromatic resins from plants of the genus *Commiphora*. The principal resin used was balm. This where we get the words "embalm" and "embalming."

The earliest embalming method called for the corpse to be wrapped in cloth and buried in charcoal and sand in an area free of humidity. But by the beginning of the New Kingdom, around 1570 B.C., embalming was such an art form that ancient Egyptian morticians competed against one another, trying to turn out the most realistic visage of life possible, often by painting the face and using other body adornments.

The embalming of ancient Egyptian royalty involved major surgery. The heart, lungs, and intestines were excised, washed in palm wine, and sealed in urns filled with alcohol and herbs.

The brain was the one organ that was discarded. It was believed to be merely a functionless mass—regarded in the same way as we regard fat.

Myrrh, other fragrant resins and oils, and specially prepared perfumes were poured into the eviscerated body cavity, and the incision was stitched closed. The body was packed in saltpeter, an astringent, for two months, then removed, soaked in wine, swathed snugly in

APPROXIMATELY 10 percent of all people who ever lived on the face of the Earth are alive today.

cotton bandages, dipped in a porridgelike antibacterial paste, lowered into an ornate coffin, and housed in a sepulcher.

The ancient Egyptians were such excellent surgeons that brain surgery was not beyond their scope. They learned such skills in order to preserve the dead as much as to cure the living. By the beginning of Christianity, however, Egyptian mummification and embalming techniques were lost.

No human society has ever achieved immortality, but the Egyptians certainly tried.[33]

Is it possible to be buried alive?

That's not possible today. These days, people are embalmed, so being buried alive is impossible. However, this was not always the case. In the 1790s, loud horns and electric shocks were used to make sure the dead were really dead. At that time, there were lots of horror stories about people being accidentally buried alive. Very few were actually true, but that didn't stop people from ordering bells and other devices to be put on their graves, so they could sound the alarm if there had been a terrible mistake. A common way to check whether someone was dead was to put a mirror in front of their mouth. The slightest breath would make the mirror go cloudy, telling you the person was still alive.

afterword

Of course, there are plenty of OBQs—Odd Body Questions—that simply aren't answered in this book. We're sorry if your favorite was left out. But don't despair. We think we have a solution. You can have your OBQ answered in the *next* OB book. That's right. If the demand is there, we'll write another OB book to follow this one.

Send your OBQ to the author:

Dr. Stephen Juan
c/o Andrews McMeel Publishing
4520 Main Street
Kansas City, MO 64111

s.juan@edfac.usyd.edu.au

notes

chapter 1: beginnings

1. Kimbel, W., Johanson, D., and Rak, Y. (1994) The First Skull and Other New Discoveries of *Australopithecus Afarensis* at Hadar, Ethiopia. *Nature* 368:449–51.

2. White, T., Suwa, G., and Asfaw, B. (1994) Australopithecus Ramidus: A New Species of Early Hominid from Aramis, Ethiopia. *Nature* 371:306–12.

3. Wong, K. (2003) An Ancestor to Call Our Own. *Scientific American,* January, pp. 54–63.

4. Chamberlain, D. (1994) The Sentient Prenate: What Every Parent Should Know. *Pre- and Perinatal Psychology Journal* 9:9–31, pp. 12–13.

5. Birnholz, J., Stephens, J., and Faria, M. (1978) Fetal Movement Patterns: A Possible Means of Defining Neurologic Development Milestones in Utero. *American Journal of Roentology* 130:537–40.

6. Liley, A. (1972) The Foetus as a Personality. *Australian and New Zealand Journal of Psychiatry* 6:99–105.

7. Birnholz, J. (1981) The Development of Human Fetal Eye Movement Patterns. *Science* 213:679–81.

8. Anand, K., and Hickey, P. (1987) Pain and Its Effects in the Human Neonate and Fetus. *New England Journal of Medicine* 317:1321–29.

9. Hooker, D. (1969) *The Prenatal Origins of Behavior.* New York: Hafner.

10. Juan, S. (1989) For Baby, a Touch of Love Will Do Nicely. *Sydney Morning Herald,* 14 December, p. 17.

11. Morse, M. (1994) Life Inside the Womb. *Parents,* December, p. 76.

12. Juan, S. (1989) How We've Been Thick-Skinned About Touching. *Sydney Morning Herald,* 9 November, p. 13.

13. Peleg, D., and Goldman, J. (1980) Fetal Heart Rate Acceleration in Response to Light Stimulation as a Clinical Measure of Fetal Wellbeing: A Preliminary Report. *Journal of Perinatal Medicine* 8:38–41.

14. Juan, S. (1989) Baby's Eye-View Described. *Eye Care Australia,* October, p. 8.

15. Shahidullah, S., and Hepper, P. (1992) Hearing in the Fetus: Prenatal Detection of Deafness. *International Journal of Prenatal and Perinatal Studies* 4:235–40.

16. Chamberlain, D. (1998) *The Mind of Your Newborn Baby.* Berkeley, Calif.: North Atlantic Books.

17. Chamberlain, D. (1994) The Sentient Prenate: What Every Parent Should Know. *Pre- and Perinatal Psychology Journal* 9:9–31, pp. 12–13.

18. Wertheimer, M. (1961) Psychomotor Coordination of Auditory and Visual Space at Birth. *Science* 134:1692–93.

19. Morse, M. (1994) Life Inside the Womb. *Parents,* December, p. 76.

20. Disher, D. (1934) The Reactions of Newborn Infants to Chemical Stimuli Administered Nasally. In Dockeray, F., ed. *Studies in Infant Behavior.* Columbus: Ohio State University Press, pp. 1–52.

21. Steiner, J. (1979) Human Facial Expression in Response to Taste and Smell Stimulation. In Reese, H., and Lipsitt, L., eds. *Advances in Child Development and Behavior.* New York: Academic Press, pp. 257–95.

22. Maurer, D., and Maurer, C. (1988) *The World of the Newborn.* New York: Basic Books, pp. 86–87.

23. Juan, S. (1989) How Does a Baby Smell? Pretty Well, by All Reports. *Sydney Morning Herald,* 16 November, p. 16.

24. Tatzer, E., Schubert, M., Timischl, W., and Simbruner, G. (1985) Discrimination of Taste and Preference for Sweet in Premature Babies. *Early Human Development* 12:23–30.

25. Morse, M. (1994) Life Inside the Womb. *Parents,* December, p. 76.

26. DeSnoo, K. (1937) Das trinkende Kind im Uterus. *Monatsschrift für Geburtshilfe und Gynaekologie* 105:88–97.

27. Freed, F. (1927) Report of a Case of Vagitus Uterinus. *American Journal of Obstetrics and Gynecology* 14:87–89.

28. Juan, S. (1991) About Face. *24 Hours,* January, p. 20.

29. Hepper, P., Shaidullah, S., and White, R. (1990) Origins of Fetal Handedness. *Nature* 347:431.

30. Luria, A. (1969) *The Mind of a Mnemonist.* New York: Avon Books.

31. Hoffmann, R. (1978) Developmental Changes in Human Infant Visual-Evoked Potentials to Patterned Stimuli Recorded at Different Scalp Locations. *Child Development* 49:110–18.

32. Maurer, D., and Maurer, C. (1988) *The World of the Newborn.* New York: Basic Books, pp. 65–66.

33. Lewkowicz, D., and Turkewitz, G. (1980) Cross-Modal Equivalence in Early Infancy: Auditory-Visual Intensity Matching. *Developmental Psychology* 16:597–607.

34. Maurer, D., and Maurer, C. (1988) *The World of the Newborn.* New York: Basic Books, pp. 66–67.

35. Cytowic, R. (1989) *Synesthesia: A Union of the Senses.* New York: Springer-Verlag.

36. Cytowic, R. (1993) *The Man Who Tasted Shapes.* New York: Jeremy Tarcher/Putnam.

37. Baron-Cohen, S., Wyke, M., and Binnie, C. (1987) Hearing Words and Seeing Colours: An Experimental Investigation of a Case of Synaesthesia. *Perception* 16:761–67.

38. Motluk, A. (1994) The Sweet Smell of Purple. *New Scientist,* 13 August, pp. 32–37.

39. Juan, S. (1989) When Commonsense Is Nonsense. *Sydney Morning Herald,* 5 October, p. 13.

40. Abrams, M. (2003) Can You See with Your Tongue? *Discover,* June, pp. 52–56.

41. Hansen, J. (1980) "Blindsight"—Seeing Without Realizing That You Can See. *Science Digest,* January, p. 14.

42. Horgan, J. (1994) Can Science Explain Consciousness? *Scientific American,* July, pp. 88–94, p. 91.

43. Caudill, M. (1992) *In Our Own Image: Building an Artificial Person.* Melbourne: Oxford University Press.

44. Martino, J. (1992) Robotic Luggage Carrier. *The Futurist,* July–August, p. 6.

45. Tarlow, P., and Muehsam, M. (1992) Wide Horizons: Travel and Tourism in the Coming Decades. *The Futurist,* November–December, pp. 28–32.

46. Martino, J. (1992) Artificial Muscle. *The Futurist,* November–December, p. 8.

47. Juan, S. (1992) Intelligent Androids: The Next Generation. *Sydney Morning Herald,* 10 December, p. 16.

48. Rosenfeld, A. (1992) The Medical Story of the Century. *Longevity,* May, pp. 42–53, p. 52.

49. Rosenfeld, A. (1992) The Medical Story of the Century. *Longevity,* May, pp. 42–53, p. 53.

50. Juan, S. (1992) Soon the Old May Be Able to Grow Younger. *Sydney Morning Herald,* 21 May, p. 12.

51. Kahn, C. (1995) Long-Life Forecast: Health Predictions You Can Use Now. *Longevity,* March, p. 14.

52. Perls, T. (1995) The Oldest Old. *Scientific American,* January, pp. 70–75, p. 70.

chapter 2: the brain

1. Sturge-Weber Foundation. (1994) *Sturge-Weber Syndrome: An Overview for Physicians and Families* (Video). Aurora, Colo.: Sturge-Weber Foundation.

2. McCutcheon, M. (1989) *The Compass in Your Nose.* Melbourne: Schwartz & Wilkinson, pp. 64–65.

3. United Press International. (1990) Study Finds More Southpaws Among Gays and Lesbians. *San Francisco Chronicle,* 26 July, p. B3.

4. Schmeck, H. (1988) Q & A. *New York Times,* 7 June, p. B7.

5. Juan, S. (1993) Get It Right or Die Trying. *Sydney Morning Herald,* 4 August, p. 14.

6. McMonnies, C. (1990) Left-Right Discrimination in Adults. *Clinical and Experimental Optometry* 73:155–58.

7. McMonnies, C. (1992) Visual-Spatial Discrimination and Mirror Letter Reversals in Reading. *Journal of the American Optometric Association* 63:698–704.

8. Schmeck, H. (1988) Q & A. *New York Times,* 23 February, p. Y19.

9. Browne, M. (1993) "Handedness" Seen in Nature, Long Before Hands. *New York Times,* 15 June, pp. B5, B7.

10. Halpern, D., and Coren, S. (1988) Longevity and Handedness. *Nature* 333:213.

11. Anderson, M. (1989) Lateral Preference and Longevity. *Nature* 341:112.

12. United Press International. (1989) Lefties Live Longer, Newest Study Finds. *San Francisco Chronicle,* 14 September, p. B3.

13. Stroh, M. (1992) It's Risky to Be a Lefty. *Science News,* 23 May, p. 351.

14. Bristow, D. (1994) Left in Doubt. *New Scientist,* 28 May, p. 65.

15. Porac, C., and Friesen, I. (2000) Hand Preference Side and Its Relation to Hand Preference Switch History Among Old and Oldest-Old Adults. *Developmental Neuropsychology* 17:2:225–39.

16. Christman, S., and Propper, R. (2001) Superior Episodic Memory Is Associated with Interhemispheric Processing. *Neuropsychology* 15:4:607–16.

17. Coffin, G. (1987) Asymmetry of the Human Head: Clinical Observations. *Clinical Pediatrics* 25:230–32.

18. Juan, S. (1991) Sharing Memories May Be as Simple as Sharing a Needle. *Sydney Morning Herald,* 9 February, p. 13.

19. Stock, G. (1993) *Metaman: The Merging of Humans and Machines into a Global Superorganism.* New York: Simon & Schuster.

20. Moravec, H. (1988) *Mind Children.* Cambridge, Mass.: Harvard University Press, pp. 22–23.

21. Juan, S. (1994) And You Thought Computer Crashes Were a Headache Now . . . *Sydney Morning Herald,* 23 March, p. 13.

22. Gorrell, C. (2003) 1890s State of Mind. *Psychology Today,* April, p. 96.

23. Graham-Rowe, D. (2003) The World's First Brain Prosthesis. *New Scientist,* 15 March, pp. 4–5.

chapter 3: the head

1. Brennan, T., Funk, S., and Frothingham, T. (1985) Disproportionate Intra-Uterine Head Growth and Developmental Outcome. *Developmental Medicine and Child Neurology* 27:746–50.

2. Elliman, A. M., Bryan, E., Elliman, A. D., and Starte, D. (1986) Narrow Heads of Preterm Infants—Do They Matter? *Developmental Medicine and Child Neurology* 28:745–48.

3. Edell, D. (1986) Hole in the Head. *People's Medical Journal,* August, p. 8.

4. Smith, W. (1964) *Bore Hole.* New York: Tribune Press.

5. Juan, S. (1987) Human Skull Theories That Could Turn Heads. *Times on Sunday,* 5 April, p. 31.

6. Fleming, C. (1988) If We Can Keep a Severed Head Alive. . . . *British Medical Journal* 297:1048.

7. Fleming, C. (1988) *If We Can Keep a Severed Head Alive.* St. Louis: Polinym Press.

8. White, R., Wolin, L., Massopust, L., Taslitz, N., and Verdura, J. (1971) Cephalic Exchange Transplantation in the Monkey. *Surgery* 70:135–39.

9. Hamblin, T. (1988) Nipping in the Bud. *British Medical Journal* 297:629.

10. Juan, S. (1990) There's Life in the Old Head. *Sydney Morning Herald,* 26 April, p. 15.

11. Adebonojo, F. (1991) Infant Head Shaping. *Journal of the American Medical Association* 265:1179.

12. Epstein, F., Hochwald, G., and Ransohoff, J. (1973) Neonatal Hydrocephalus Treated by Compressive Head Wrapping. *The Lancet* 1:634–36.

13. Gould, S. (1981) *The Mismeasure of Man.* London: W. W. Norton.

14. Juan, S. (1992) Had a Facelift? Then Try a Reshaped Skull. *Sydney Morning Herald,* 5 November, p. 12.

15. Sugar, O. (1971) Head Shrinking. *Journal of the American Medical Association* 216:117–20.

16. Karsten, R. (1935) The Head Hunters of Western Amazonas. *Societas Scientiarum Fennica, Commentationes Humanarum Litterarum* 7:1–588.

17. Stirling, M. (1938) Historical and Ethnological Material on the Jivaro Indians. *Bulletins of the Bureau of American Ethnology* 117:1–148.

18. Leavesley, J. (1994) Headshrinking as an Art Form. *Australian Doctor,* 9 September, p. 81.

19. Gould, S. (1977) *Ever Since Darwin.* New York: W. W. Norton, p. 63.

chapter 4: the eyes

1. Ferrer, H. (1928) Voluntary Propulsion of Both Eyeballs. *American Journal of Ophthalmology* 11:883.

2. Smith, J. (1932) Voluntary Propulsion of Both Eyeballs. *Journal of the American Medical Association* 98:398.

3. Berman, B. (1966) Voluntary Propulsion of the Eyeball: The Double Whammy Syndrome. *Archives of Internal Medicine* 117:648–51.

4. Borgquist, A. (1906) Crying. *American Journal of Psychology* 27:149–205.

5. Levoy, G. (1988) Crying It Out. *San Jose Mercury News,* 2 November, pp. 1F, 3F.

6. Carey, B. (1988) Vital Statistics. *Hippocrates,* September–October, p. 16.

7. Rymer, R. (1989) Why Do Women Cry More Than Men? *Hippocrates,* January–February, p. 100.

8. Juan, S. (1988) Why a Good Cry Can Make You Feel Better. *Sydney Morning Herald,* 28 October, p. 13.

9. Juan, S. (1989) How to Spot Some Signs of Aging. *Sydney Morning Herald,* 13 February, p. 15.

10. Troiano, L. (1989) The Wink of an Eye. *American Health,* November, p. 40.

11. Albert, B. (1989) What Causes Bags and Rings Under the Eyes? *University of California, Berkeley Wellness Letter,* June, p. 8.

12. Albert, B. (1989) What's a Sty? *University of California, Berkeley Wellness Letter,* July, p. 8.

13. Kiefer, J. (1988) What Are Those Spots That Sometimes Float Before My Eyes? *Hippocrates,* November–December, p. 116.

14. Bower, B. (1991) Vision System Puts Eyesight in Blind Spots. *Science News,* 27 April, p. 262.

15. Lipkin, R. (1993) Focusing the Soul's Fuzzy Window. *Science News,* 11 September, p. 172.

16. Wright, S. (1990) How Carrots Can Aid Eyesight. *San Jose Mercury News,* 13 February, p. C1.

17. Fackelmann, K. (1993) Nutrients May Prevent Blinking Disease. *Science News,* 12 November, p. 310.

18. Fackelmann, K. (1993) Nocturnal Risks for the Eyes. *Science News,* 8 May, p. 302.

19. Franklin, D. (1992) Tuning the Kids In. *In Health,* December–January, p. 40.

20. Siebers, T. (1983) The Mirror of Medusa. Berkeley and Los Angeles: University of California Press.

21. Maloney, C. (ed.) (1976) *The Evil Eye.* New York: Columbia University Press.

22. Juan, S. (1991) People Still Feel Deadly Eye's Power. *Sydney Morning Herald,* 8 August, p. 12.

23. Juan, S. (1989) How Soon Can You Tell Colour? *Eye Care Australia,* September, p. 12.

chapter 5: the nose, ears, and mouth

1. Blakeslee, S. (1988) Q & A. *New York Times,* 12 January, p. Y19.

2. Provine, R., Hamernik, H., and Curchack, B. (1987) Yawning: Relation to Sleeping and Stretching in Humans. *Ethology* 76:152–60.

3. Adams, A. (1998) The Big Yawn. *New Scientist,* 19 December, p. 72.

4. Boyce, N. (1999) Yawning Shows We're Just Big Babies. *New Scientist,* 10 April, p. 72.

5. Juan, S. (1999) The Psychology of Smell. "Tempo." *Sun-Herald* (Sydney), 4 April, p. 13.

6. Wysocki, C. (2002) Do People Lose Their Senses of Smell and Taste as They Age? *Scientific American,* June, p. 107.

7. Rensberger, B. (1992) This Story Is Bound to Make You Yawn. *Washington Post,* 12 November, pp. D1, D6.

8. McCarthy, P. (1987) Yawning to Breathe Free? *Psychology Today,* February, p. 9.

9. Daquin, G., Micallef, J., and Blin, O. (2001) Yawning. *Sleep Medicine Review* 5:4:299–312.

10. Juan, S. (1988) Why One Yawn Can Lead to Another. *Sydney Morning Herald,* 14 July, p. 16.

11. Juan, S. (1988) When It's All in the Ear of the Beholder. *Sydney Morning Herald,* 9 August, p. 17.

12. Blakeslee, S. (1988) Q & A. *New York Times,* 8 December, p. Y26.

13. Andersson, M. (2003) Sink Stoppage. *New Scientist,* 22 March, p. 65.

14. Davis, L. (1991) What's That Ringing in My Ear? *In Health,* September–October, p. 100.

15. Margaretten-Ohring, J. (1990) Tinnitus: When Your Ears Ring and Ring. *University of California, Berkeley Wellness Letter,* March, p. 7.

16. Juan, S. (1988) When All That Hissing and Buzzing Becomes Too Much. *Sydney Morning Herald,* 24 August, p. 17.

17. Cooper, M. (1992) *Winning with Your Voice.* Hollywood, Fla.: Fell Publishers.

18. Moody, L. (1992) What Your Voice Says About You. *Los Angeles Daily News,* 13 May, pp. D2–D3.

19. Zamichow, N. (1992) The Urge to "Um." *Los Angeles Times,* 29 April, p. B4.

20. Juan, S. (1992) Sick of the Sound of Your Own Voice. *Sydney Morning Herald,* 12 November, p. 12.

21. Ray, C. (1991) Q & A. *New York Times,* 23 July, p. Y18.

22. Carbary, T., Patterson, J., and Snyder, P. (2000) Foreign Accent Syndrome Following a Catastrophic Second Injury: MRI Correlates, Linguistic and Voice Pattern Analyses. *Brain and Cognition* 43:1–3:78–85.

23. Reeves, R., and Norton, J. (2001) Foreign Accent-Like Syndrome During Psychotic Exacerbations. *Neuropsychiatry, Neuropsychology, Behavioral Neurology* 14:2:135–38.

24. Gurd, J., Coleman, J., Costello, A., and Marshall, J. (2001) Organic or Functional? A New Case of Foreign Accent Syndrome. *Cortex* 37:5:715–18.

25. Blumstein, S., Alexander, M., Ryalls, J., Katz, W., and Dworetzky, B. (1987) On the Nature of the Foreign Accent Syndrome: A Case Study. *Brain and Language* 31:2:215–44.

26. Kurowski, K., Blumstein, S., and Alexander, M. (1996) The Foreign Accent Syndrome: A Reconsideration. *Brain and Language* 54:1:1–25.

27. Chui, G. (1989) The Bulbous Red Nose. *San Jose Mercury News,* 20 June, p. 2C.

28. Goldsmith, M. (1989) New Topical Therapy for Acne Rosacea Offers Conspicuous Improvement, No Systemic Effects. *Journal of the American Medical Association* 261:2014–15.

29. Juan, S. (1989) A Cure for Rudolph. *Sydney Morning Herald,* 7 September, p. 12.

30. Xenakis, A. (1993) *Why Doesn't My Funny Bone Make Me Laugh?* New York: Villard Books, pp. 85–86.

31. Blakeslee, S. (1987) Q & A. *New York Times,* 3 October, p. Y16.

32. Juan, S. (1993) A Hard-to-Swallow Ailment. *Sydney Morning Herald,* 24 November, p. 13.

33. Ray, C. (1992) Q & A. *New York Times,* 23 June, p. B8.

34. Hart, C. (1994) Light Sneeze. *New Scientist,* 25 June, p. 65.

35. Eccles, R. (1994) Light Sneeze. *New Scientist,* 21 May, p. 57.

36. Ray, C. (2003) Earlobe Crease. *New York Times,* 29 April, p. D2.

37. Leahy, S. (2003) Rough Sense of Smell. *New Scientist,* 12 April, p. 26.

chapter 6: the skin

1. Syme, S. (1988) The Itch That Dare Not Speak Its Name. *University of California, Berkeley Wellness Letter,* March, pp. 6–7.

2. Zuger, A. (2003) The Mystery of Itch, the Joy of Scratch. *New York Times,* 1 July, p. D1, D5.

3. Juan, S. (1988) Irritating Mystery of Itching. *Sydney Morning Herald,* 12 September, p. 17.

4. Halpern, D., Blake, R., and Hillenbrand, J. (1986) Psychoacoustics of a Chilling Sound. *Perception and Psychophysics* 39:77–80.

5. Juan, S. (1988) How a Call from the Past Can Give You the Shivers. *Sydney Morning Herald,* 5 October, p. 17.

6. Schmeck, H. (1987) Q & A. *New York Times,* 17 February, p. Y19.

7. Juan, S. (1988) Thin-Skinned, Lots of Exposure and Likely to Wrinkle. *Sydney Morning Herald,* 14 November, p. 21.

8. Juan, S. (1994) Wrinkles? That's Stretching It a Bit. . . . *Sydney Morning Herald,* 8 June, p. 13.

9. Blakeslee, S. (1988) Q & A. *New York Times,* 23 February, p. Y19.

10. Brewer, S. (1994) Fountain of Youth for Aging Skin. *Longevity,* January, p. 5.

11. Xenakis, A. (1993) *Why Doesn't My Funny Bone Make Me Laugh?* New York: Villard Books, pp. 140–42.

12. Noll, R. (1994) Hypnotherapy for Warts in Children and Adolescents. *Journal of Developmental and Behavioral Pediatrics* 15:170–73, p. 170.

13. Juan, S. (1994) The War to End All Warts. *Sydney Morning Herald,* 22 June, p. 11.

14. Fackelmann, K. (1992) Genital-Warts Virus Linked to Penile Cancer. *Science News,* 23 May, p. 342.

15. Reinisch, J. (1992) Genital Warts Caused by Virus. *San Francisco Chronicle,* 23 June, p. B5.

16. Siegel, D., and McDaniel, S. (1991) The Frog Prince: Tale and Toxicology. *American Journal of Orthopsychiatry* 61:558–62, p. 558.

17. Juan, S. (1992) Prince Charming Was Just a Hallucination. *Sydney Morning Herald,* 7 May, p. 14.

18. Juan, S. (1989) When Your Hair Stands On End. *Sydney Morning Herald,* 25 January, p. 13.

19. Staff of the Royal Children's Hospital, Melbourne. (1992) *Paediatric Handbook.* 4th edition. Melbourne: Blackwell Scientific Publications, p. 159.

20. Blakeslee, S. (1987) Q & A. *New York Times,* 31 March, p. Y21.

21. Juan, S. (1989) It's Just a Question of Degrees. *Sydney Morning Herald,* 24 March, p. 10.

22. Wilford, J. (1988) Q & A. *New York Times,* 23 August, p. B8.

23. Juan, S. (1990) Why Men Have Nipples. *Sydney Morning Herald,* 31 December, p. 11.

24. Diamond, J. (1995) Father's Milk. *Discover,* February, pp. 82–87, p. 82.

25. Flinn, J. (1990) Are You Blushing Yet? *San Francisco Chronicle,* 28 January, pp. E17–E18.

26. Scott, J. (1990) A Blush Means "I'm Sorry." *Los Angeles Times,* 26 January, pp. D1–D4.

27. Leary, M., Britt, T., Cutlip, W., and Templeton, J. (1992) Social Blushing. *Psychological Bulletin* 112:3:446–60.

28. Gerlach, A., Wilhelm, F., Gruber, K., and Roth, W. (2001) Blushing and Physiological Arousability in Social Phobia. *Journal of Abnormal Psychology* 110:2:247–58.

29. Juan, S. (1992) Blushing: Why We Do It. *Sydney Morning Herald,* 4 February, p. 10.

30. Curtis, J., and Lawrence, K. (1994) Comes in Handy. *New Scientist,* 6 July, p. 65.

31. Nordlund, J. (1992) *Guidelines for the Treatment of Patients with Vitiligo.* Tyler, Tex.: National Vitiligo Foundation, p. 2.

32. Barclay, L. (2003) Tacrolimus Helpful for Vitiligo Treatment. *Medscape Medical News,* 28 May, pp. 1–4.

33. Juan, S. (1993) A Black and White Issue. *Sydney Morning Herald,* 17 February, p. 10.

34. Children's Hospital Medical Center of Cincinnati. (2003) Study Discovers Key to Baby-Like Skin (Press Release), 7 May, pp. 1–2.

chapter 7: the hair and nails

1. Barnhurst, B. (1995) Grey Matter. *New Scientist,* 4 February, p. 57.

2. Parachini, A. (1987) Root of Graying Remains a Mystery for Scientists. *Los Angeles Times,* 13 October, pp. 1E, 3E.

3. Juan, S. (1988) Grey Hair: Just an Optical Illusion. *Sydney Morning Herald,* 4 November, p. 17.

4. Juan, S. (1989) Fact That Will Make Most of Your Hair Curl. *Sydney Morning Herald,* 22 February, p. 19.

5. Addison, W. (1989) Beardedness as a Factor in Perceived Masculinity. *Perceptual and Motor Skills* 68:921–22, p. 921.

6. Muller, S. (1990) Trichotillomania: A Histopathological Study in Sixty-Six Patients. *Journal of the American Academy of Dermatology* 23:56–62.

7. Christenson, G., Pyle, R., and Mitchell, J. (1991) Estimated Lifetime Prevalence of Trichotillomania in College Students. *Journal of Clinical Psychiatry* 52:415–17, p. 415.

8. Rapoport, J. (1989) *The Boy Who Couldn't Stop Washing.* New York: Dutton, pp. 152–53.

9. Juan, S. (1992) This Can Make You Tear Your Hair Out. *Sydney Morning Herald,* 6 February, p. 12.

10. Repinski, K. (1992) Thicker, Youthful Hair—Another Drug to Save It? *Longevity,* January, p. 20.

11. Samman, P., and Fenton, D. (1986) *Nails in Disease.* 4th edition. St. Louis: C. V. Mosby.

12. Scher, R., and Daniel, C. (1990) *Nails: Therapy, Diagnosis, and Surgery.* Philadelphia: W. B. Saunders.

13. Juan, S. (1993) Some Theories On Nail-Biting to Chew Over. *Sydney Morning Herald,* 28 April, p. 18.

14. Senator, G. (1991) Hirsutism. *Australian Dr. Weekly,* 29 March, pp. i–vi.

15. Mauvais-Jarvis, P., Kutten, F., and Mowszowicz, I. (1981) *Hirsutism.* New York: Springer-Verlag.

16. Judd, S., and Carter, J. (1992) The Changing Face of Hirsutism. *Medical Journal of Australia* 156:148–49.

17. Zwicker, H., and Rittmaster, R. (1993) Androsterone Sulfate: Physiology and Clincial Significance in Hirsute Women. *Journal of Clinical Endocrinology and Metabolism* 76:112–16.

18. Juan, S. (1993) The Hirsute of Excellence: A Hair Question of What's Fashionable. *Sydney Morning Herald,* 20 October, p. 15.

19. Azizzadeh, A., Moldovan, S., and Scott, B. (2002) Image of the Month: Rapunzel Syndrome. *Archives of Surgery* 137:12:1443–44.

20. Deslypere, J., Praet, M., and Verdonk, G. (1982) An Unusual Case of the Trichobezoar: The Rapunzel Syndrome. *American Journal of Gastroenterology* 77:467–70.

21. Kirk, A., Bowers, B., Moylan, J., and Meyers, W. (1988) Toothbrush Swallowing. *Archives of Surgery* 123:382–84.

22. Ott, S., Helmberger, T., and Beuers, U. (2003) Intestinal Obstruction After Ingestion of a Beer-Filled Condom at the Munich Octoberfest. *American Journal of Gastroenterology* 98:2:512–13.

23. Baskonus, I., Gokalp, A., Maralcan, G., and Sanal, I. (2002) Giant Gastric Trichobezoar. *International Journal of Clinical Practice* 56:5:399–400.

24. Williams, R. (1986) The Fascinating History of Bezoars. *Medical Journal of Australia* 145:613–14.

25. Juan, S. (1988) Why the Hair Ball's Charm Is a Hard One to Swallow. *Sydney Morning Herald,* 28 July, p. 15.

26. Juan, S. (1990) *Only Human: Why We React, How We Behave, What We Feel.* Sydney: Random House Australia, pp. 150–51.

27. Brooks, A. (1994) Shiny, Happy Hair. *New Scientist,* 6 August, p. 65.

chapter 8: the skeleton, bones, and teeth

1. Angert, E., Clements, K., and Pace, N. (1993) The Largest Bacterium. *Nature* 362:239–41.

2. Juan, S. (1993) When Big Just Keeps Getting Bigger. *Sydney Morning Herald,* 22 September, p. 15.

3. Samaras, T. (1995) Trends Toward Tallness. *The Futurist,* January–February, p. 29.

4. Juan, S. (1993) Brains But No Brawn. *Sydney Morning Herald,* 6 October, p. 17.

5. Goleman, D. (1987) Q & A. *New York Times,* 7 July, p. Y15.

6. Carey, B. (1989) How Are People Able to Predict Bad Weather by Pain in Their Joints? *Hippocrates,* January–February, p. 100.

7. Juan, S. (1989) Weather Puts Arthritics Out of Joint. *Sydney Morning Herald,* 6 July, p. 14.

8. Tributsch, H. (1982) *When the Snakes Awake: Animals and Earthquake Prediction.* Cambridge, Mass.: M.I.T. Press.

9. Raskin, D. (1990) Earthquakes: Do Animals Know? *American Health,* May, p. 102.

10. Wallace, I., Wallechinsky, D., and Wallace, A. (1990) Animals That Foretell Quakes. *San Francisco Chronicle,* 23 May, p. B3.

11. Juan, S. (1990) Can Our Bodies Warn Us of Earthquakes? *Sydney Morning Herald,* 17 May, p. 17.

12. Jones, P. (1972) Dickens' Literary Children. *Australian Paediatric Journal* 8:233–45.

13. Associated Press. (1992) Tiny Tim Had Kidney Disease, Doctor Says. *San Francisco Chronicle,* 15 December, p. 3.

14. Juan, S. (1990) What May Have Ailed Tiny Tim. *Sydney Morning Herald,* 20 December, p. 12.

15. Joyce, C., and Stover, E. (1991) *Witnesses from the Grave: The Stories Bones Tell.* Boston: Little, Brown & Company.

16. Chui, G. (1991) Bones Yield Clues to Solve Mysteries of the Ages. *San Jose Mercury News,* 5 August, pp. 1A–1B.

17. Juan, S. (1991) Secrets of Identity Easily Unlocked. *Sydney Morning Herald,* 10 October, p. 16.

18. Warne, G. (1990) Contemporary Issues in the Use of Growth Hormone. *Journal of Paediatrics and Child Health* 26:122–23.

19. Fackelmann, K. (1989) Pygmy Paradox Prompts a Short Answer. *Science News,* 8 July, p. 22.

20. Hallett, J. (1973) *Pygmy Kitabu.* New York: Random House.

21. Reuters. (1992) Last Pygmies in Danger. *Australian Dr. Weekly,* 15 May, p. 48.

22. Juan, S. (1992) The Long and Short of Life as a Pygmy. *Sydney Morning Herald,* 2 July, p. 16.

23. Edell, D. (1993) Q & A with Dr. Edell. *The Edell Health Letter,* March, p. 8.

24. Xenakis, A. (1993) *Why Doesn't My Funny Bone Make Me Laugh?* New York: Villard Books, pp. 1–2.

25. Edell, D. (1991) Q & A with Dr. Edell. *The Edell Health Letter,* March, p. 8.

26. Ezzell, C. (1991) Writer's Cramp: Literally in Your Head. *Science News,* 23 November, p. 333.

27. Rosenthal, E. (1987) Q & A. *New York Times,* 29 December, p. Y17.

28. Juan, S. (1993) Do We Really Have a "Funny Bone"? *Sydney Morning Herald,* 14 April, p. 12.

29. Feldman, S. (1940) Phantom Limbs. *American Journal of Psychology* 53:590–98.

30. Melzack, R. (1992) Phantom Limbs. *Scientific American,* April, pp. 120–26.

31. Corliss, W. (1993) *Biological Anomalies: Humans II.* Glen Arm, Md.: The Sourcebook Project, pp. 205–7.

32. Riggs, L. (1995) Men Shrink an Average of 1¼ Inches. *Bottom Line*, 1 February, p. 6.

33. Wales, J., and Dangerfield, P. (1995) Night Growth. *New Scientist*, 18 February, p. 65.

chapter 9: the inside

1. Schweiger, A., and Parducci, A. (1981) Nocebo: The Psychologic Induction of Pain. *Pavlovian Journal of Biological Science* 16:140–43.

2. Morse, J., and Morse, R. (1988) Cultural Variation in the Inference of Pain. *Journal of Cross-CulturalPsychology* 19:232–42.

3. Juan, S. (1988) How We Can Feel Pain from Experience. *Sydney Morning Herald*, 1 December, p. 16.

4. Cousins, N. (1979) *Anatomy of an Illness as Perceived by the Patient.* Toronto: Bantam Books.

5. Lang, S. (1988) Laughter Is the Best Defense. *American Health*, December, p. 42.

6. Juan, S. (1989) Heard the One About the Immune System? *Sydney Morning Herald*, 18 May, p. 15.

7. Lang, S. (1988) Laughing Matters—At Work. *American Health*, September, p. 46.

8. Peterson, C., Seligman, M., and Vaillant, G. (1988) Pessimistic Style Is a Risk Factor for Physical Illness: A Thirty-Five-Year Longitudinal Study. *Journal of Personality and Social Psychology* 55:23–27.

9. Peterson, C. (2003) Personal Communication, 30 May.

10. Juan, S. (1988) Laughing All the Way to the Top. *Sydney Morning Herald*, 22 September, p. 14.

11. Ziv, A. (1988) Teaching and Learning with Humor: Experiment and Replication. *Journal of Experimental Education* 57:5–15.

12. Dimmer, S., Carroll, J., and Wyatt, G. (1990) Uses of Humor in Psychotherapy. *Psychological Reports* 66:795–801.

13. Fry, W., and Salameh, W. (eds.) (1987) *Handbook of Humor in Psychotherapy.* New York: Pergamon Press.

14. Juan, S. (1991) Research into Humour No Laughing Matter. *Sydney Morning Herald,* 20 June, p. 14.

15. Bass, S. (1989) Wearing a Smile May Cheer You Up. *New York Times,* 1 August, p. B1.

16. Stone, J. (1990) What's in Her Smile? *American Health,* September, pp. 30–35.

17. Juan, S. (1991) About Face. *24 Hours,* January, p. 20.

18. Rohter, L. (1987) Q & A. *New York Times,* 7 July, p. Y15.

19. Juan, S. (1991) You Can't Tickle Yourself. *Sydney Morning Herald,* 9 January, p. 12.

20. Rosenthal, E. (1989) Q & A. *New York Times,* 7 November, p. B7.

21. Juan, S. (1989) Santa's Big Belly Comes in for Some Serious Prodding. *Sydney Morning Herald,* 21 December, p. 10.

22. Weiss, R. (1992) Travel Can Be Sickening; Now Scientists Know Why. *New York Times,* 28 April, p. B5.

23. Schmeck, H. (1987) Q & A. *New York Times,* 11 August, p. Y23.

24. Juan, S. (1987) Avoiding Motion Sickness. *Australian Dr. Weekly,* 30 October, p. 28.

25. Associated Press. (1988) Motion Sickness. *San Jose Mercury News,* 14 September, p. B3.

26. Livingston, K. (1988) Trying to Stop Motion Sickness. *San Francisco Chronicle*, 7 December, p. C13.

27. Juan, S. (1989) Stop the Car, Now! *Sydney Morning Herald*, 3 August, p. 10.

28. Angier, N. (1988) Q & A. *New York Times*, 5 July, p. B11.

29. Wood, C. (1976) ABO Blood Groups Related to the Selection of Human Hosts by Yellow Fever Vector. *Human Biology* 48:337–49.

30. Jorgensen, G. (1977) A Contribution to the Hypothesis of a "Little More Fitness" of Blood Group O. *Journal of Human Evolution* 6:741–54.

31. Beardmore, J., and Karimi-Booshehri, F. (1983) ABO Genes Are Differentially Distributed in Socio-Economic Groups in England. *Nature* 303:522–24.

32. Nagorka, J. (1990) Genetic Key to Blood Types Found. *Dallas Morning News*, 15 July, p. D14.

33. Juan, S. (1994) A Bloody Mystery Remains Unsolved. *Sydney Morning Herald*, 6 July, p. 11.

chapter 10: endings

1. Dolnick, E. (1989) Scared to Death. *Hippocrates*, March–April, pp. 106, 108.

2. Australian Associated Press. (1989) Doc Died Happy. *Sydney Morning Herald*, 23 May, p. 10.

3. Gould, G., and Pyle, W. (1897) *Anomalies and Curiosities of Medicine.* Philadelphia: W. B. Saunders, p. 523.

4. Corliss, W. (1992) *Biological Anomalies: Humans I.* Glen Arm, Md.: The Sourcebook Project, p. 78.

5. Michell, J., and Rickard, R. (1977) *Phenomena: A Book of Wonders.* New York: Pantheon Books, p. 24.

6. Cooper, D., and Cooper, R. (1838) On the Luminosity of the Human Subject After Death, with Remarks and Details of Experiments Made with a View of Determining the Nature of the Fuel. *Philosophical Magazine* 3:12:420–26.

7. Landy, H., Weiner, S., Corson, S., Batzer, F., and Bolognese, R. (1986) The "Vanishing Twin": Ultrasonographic Assessment of Fetal Disappearance in the First Trimester. *American Journal of Obstetrics and Gynecology* 155:14–19.

8. Juan, S. (1990) *Only Human: Why We React, How We Behave, What We Feel.* Sydney: Random House Australia, pp. 103–4.

9. Brown-Sequard, C. (1984) Attitudes After Death. *Knowledge* 6:115–18.

10. Rogo, D. (1989) *The Return from Silence.* Wellingborough: Aquarian Press, p. 24.

11. Moody, R. (1975) *Life After Life.* New York: Mockingbird Books.

12. Morse, M. (1992) *Transformed by the Light.* New York: Villard Books.

13. Ring, K. (1982) *Life After Death.* New York: Quill.

14. Ring, K. (1993) *The Omega Project.* New York: William Morrow.

15. Mauro, J. (1992) Bright Lights, Big Mystery. *Psychology Today,* July–August, pp. 54–57, 80–82, p. 82.

16. Juan, S. (1987) Memories Brought Back from Death. *The Times on Sunday,* 1 March, p. 32.

17. Stack, S., and Lester, D. Born Under a Bad Sign? Astrological Sign and Suicide Ideation. *Perceptual and Motor Skills* 66:461–62, p. 461.

18. Juan, S. (1989) A Tendency Towards Suicide May Just Be a Bad Sign from the Sun. *Sydney Morning Herald,* 23 November, p. 17.

19. Reuters. (1990) Boy Keeps Bursting into Flames. *Sydney Morning Herald,* 2 May, p. 19.

20. Randles, J., and Hough, P. (1989) *Death by Supernatural Causes.* London: Grafton.

21. Knight, B. (1989) Rainy-Day Read. *New Scientist,* 28 January, p. 74.

22. Juan, S. (1989) Combustion Theory Goes Up in Smoke. *Sydney Morning Herald,* 8 March, p. 17.

23. Lieber, A. (1978) Human Aggression and the Lunar Synodic Cycle. *Journal of Clinical Psychiatry* 39:385–93.

24. Lester, D. (1979) Temporal Variation in Suicide and Homocide. *American Journal of Epidemiology* 109:517–20.

25. Little, G., Bowers, R., and Little, L. (1987) Geophysical Variables and Behavior. *Perceptual and Motor Skills* 64:1212.

26. Coates, W., Jehle, D., and Cottington, E. (1989) Trauma and the Full Moon: A Waning Theory. *Annals of Emergency Medicine* 18:763–65.

27. Juan, S. (1990) Does a Full Moon Turn Us into Lovers or Lunatics? *Sydney Morning Herald,* 22 February, p. 15.

28. Lipa, B., Sturrock, P., and Rogot, F. (1976) Search for Correlation Between Geomagnetic Disturbances and Mortality. *Nation* 259:302–4.

29. Tatsanaviavat, P., Chiravatkul, A., Klungboonkrong, V., Chaisiri, S., Jarerntanyaruk, L., Munger, R., and Saowakontha, S. (1992) Sudden and Unexplained Deaths in Sleep (Laitai) of Young Men in Rural Northeastern Thailand. *International Journal of Epidemiology* 21:904–10.

30. Lown, B., DeSilva, R., and Lenson, R. (1978) Roles of Psychologic Stress and Autonomic Nervous System Changes in Provocation of Ventricular Premature Complexes. *American Journal of Cardiology* 41:979–85 (Afterword).

31. Juan, S. (1993) Asian Men Victims of Sudden Death. *Sydney Morning Herald,* 21 January, p. 12.

32. Juan, S. (1991) Exhuming the Truth About Zombies Can Be Sickening. *Sydney Morning Herald,* 16 May, p. 15.

33. Juan, S. (1991) An Afterlife Mint in Mummification. *Sydney Morning Herald,* 12 September, p. 12.

index

Note: Page numbers in italics indicate boxed text.

about the author

Dr. Stephen Juan, the "Wizard of Odds," is a behavioral scientist, educator, journalist, and author. An anthropologist by training and one of the world's best communicators of research, Dr. Juan was born in the Napa Valley in California. He received his BA, MA, and PhD from the University of California at Berkeley and has taught at the University of Sydney in Australia for more than twenty-five years. Currently he divides his time between homes in Sydney, Canberra, and California. A lively and popular speaker, Dr. Juan appears regularly on Australian news, current affairs, and lifestyle television and radio programs. His comments frequently appear in the Australian press. Dr. Juan addresses any and all topics having to do with being a human being. The body, the brain, the future of the human race—all things human are the focus of his attention. In his exceedingly diverse career, Dr. Juan is a pioneer in the field of developmental anthropology, a leader in science education, and the founding editor of the award-winning Australian magazine *Better Parenting*. He has also served as the resident human-behavior commentator on the Australian version of the *Big Brother* television show. As a professional speaker, Dr. Juan has given after-dinner talks and convention addresses around the world. He has been honored in Australia and overseas for his writing and other public education work. The American Medical Association and the U.S. National Association of Physician Broadcasters are among the international organizations honoring him. His books have been translated into fourteen languages.